AF661370

LIMNOLOGY

by

S.C. AGRAWAL
Department of Botany,
University of Allahabad,
Allahabad

A.P.H. PUBLISHING CORPORATION
ANSARI ROAD, DARYA GANJ
NEW DELHI-110 002

Published by
S.B. Nangia
A P H Publishing Corporation
4435-36/7, Ansari Road, Daryaganj
New Delhi 110002
Ph.: 23274050
E-mail : aphbooks@gmail.com

2026

© Author

Rs. 2495/-

Printed at
Balaji Offset
Navin Shahdara, Delhi 110032

PREFACE

This book is intended to fill the need for a text for students who need an introduction to morphology, ecology and related field of algae, vascular plants and bacteria occurring in lakes. It will be useful to both graduate and undergraduate students. After an introduction outlining the general characteristics of lakes, the book goes on to look at the importance of the algae, higher plants and bacteria occurring in lakes. The effects of different physical and chemical factors such as light, temperature, waves, nutrients, gases, pH etc. on the growth of lake flora are given. The importance of algae in symbiotic and parasitic relationships and their considerable significance to animal grazers in aquatic food chains are discussed. The relationships of lake fora to eutrophication and pollution of water are also dealt with.

I am indebted to my teachers, particularly my esteemed guide late Professor Y.S.R.K. Sarma, Department of Botany, Banaras Hindu University, who has been a source of continuous encouragement and inspiration to me. I would also like to thank all my students, graduate and undergraduate, who have been a continuous source of ideas and stimulation to me.

Allahabad, October 1998 **S.C. AGRAWAL**

CONTENTS

I

General Characteristics of Lakes

1.1 DEFINITION

The prefix 'limno' in the world limnology refers to the lakes. The lakes are quiet large bodies of freshwater usually deep enough that their bed lie much beyond the photosynthetic zone. The lakes occupy a relatively small portion of the earth surface and majority of them are situated in temperate and subarctic regions. Due to the large surface area or expansion of the open water of most of the lakes, the wind induced water turbulence is more pronounce in them than in the ponds. In ponds, the light penetrates to all depth, encouraging the growth of rooted plants nearly anywhere, but in the lakes the benthic flora develops as a girdle around the periphery leaving the centre free for plankton.

Because of the relatively shallow depth and small volume of ponds, both the dissolved oxygen content and the temperature of their waters are often uniform at all depth at a particular time, but can undergo pronounce diurnal variations. The dissolved oxygen content of a pond or a lake water is usually higher during day time (due to photosynthetic release) than in the night, and sometimes it can be so low at night that it may create problem to fish respiration. The dissolved oxygen content of pond water varies widely from supersaturation at day time when there is excessive photosynthesis to near depletion at night when respiration and decomposition predominates. However, both the dissolved oxygen content and the temperature of a lake water will continuously decrease at various levels with an increase in the depth and will also change slowly with a change in the day time and season.

1.2 DEPTH OF LAKES

Most lakes have maximum depth of less than 30 metres, but some may be much deeper, *e.g.,* Great lakes of North America may vary from 64 to 393 metres in depth and Crater lake in Oregon is 608 metres deep. The lakes can be classified into three categories depending upon their depth.

(i) *Lakes of first order* : In general, these lakes are over 90 metres in depth and their bottom water remains at or near 4°C throughout year, although one or two circulation periods are possible.

(ii) *Lakes of second order* : They are those lakes which have depth varying between about 8 to 90 metres and their bottom water temperature is more than 4°C during summer and they have one or two regular circulation periods during a year.

(iii) *Lakes of third order* : They are essentially large ponds and are less than 8 metres of depth. They do not develop extensive thermal stratification and undergo more or less continuous water circulation.

The life span of a lake is measured in terms of few hundred years to hundreds of thousands of years depending upon its depth. All lakes and ponds undergo slow but continuous filling-in process which eventually led them to extinct This is completely a natural and universal phenomenon. The lake basin is filled-in either through shoreline erosion by wind or wave action, sediment deposition washed in from watershed as a result of soil erosion or accumulation of remains of dead plants and animals of the water as well as of the surrounding land. The dead leaves blown in the water settle to the bottom and contribute to the accumulating debris. The calcareous molluskan shells persist under non-acidic conditions and develop thick deposits on the lake bottom.

1.3 RETENTION AND REPLACEMENT OF WATER IN LAKES

Very large lakes are generally characterized by having a slow exchange of water and in them the inflow and outflow of water is very small as compared to their volumes, but small lakes and ponds undergo rapid replacement of water, *e.g.* the whole water mass of a very large lake Baikal is replaced over a very long period of approximately 500 years, but that of an eleven mile long lake Windermere is replaced only in nine months.

The rate of water replacement of a lake also depends upon the season, *e.g.* during monsoon rains, the lake water may flush out rapidly over a short period of only few days and may remain static thereafter for long time, but on the other hand, in temperate regions the water inflow is very small during soil freezing, but will increase during spring thawing.

In the summer, the bulk of algal cells are concentrated in the shallow epilimnion, and if any new water enters into the lake at the water surface, it will dilute the population and will bring some new nutrients in the epilimnion and will stimulate the growth of phytoplankton, but the surface outflow of water from a lake results into a loss of phytoplankton.

Depending upon the mode of formation of a lake in the remote past, it may be either of drainage or seepage type.

(i) *Drainage type* : These are those lakes which are either stream or spring fed and have a water outlet at least during part of the year. The water drainage prevent the lake from achieving a constantly high nutrients concentrations.

(ii) *Seepage type* : They are land-locked lakes and are fed only by seepage and have no outlet. They are doomed consequently to extinction. They usually have a sandy bottom.

1.4 ORIGIN OF LAKES

Lakes are relatively recent features and can be classified into following types on basis of their origin.

(i) *Tectonic lakes* : These lakes are created as a result of some kind of deformation in the earth crust associated with earth movement during earthquakes and faults. They have rocky basins and shorelines and the materials washing in from the surrounding watershed tends to be infertile. Initial productivities are low in these lakes because of great depth and lack of nutrients, but as they age they accumulate nutrients and have richer substrates. Most of the tropical lakes are tectonic lakes.

(ii) *Volcanic lakes* : They are formed when mouth of an old volcanos is filled with water. In many respect the conditions for life in these lakes are much like those in tectonic lakes. The substrates are hard, coarse and inhospitable to most rooted plants. The lakes are deep with steep sloping sides and have restricted biota. With time, finer sediments and plant nutrients accumulate and these lakes become more productive. Some Japanese and Philippine lakes are volcanic lakes.

(iii) *Glacial lakes* : Glaciers can also form lakes and ponds. Glaciers gouge out a deep basin as they advances and then to leave a natural dam across the mouth of the basin when they retreats. The materials incorporated into glaciers are released when ice melted. The lakes formed in this manner have many of the biological characteristics of tectonic and volcanic lakes. Most of the temperate lakes are glacial lakes.

(iv) *Solution basins* : Numerous small lakes are formed in the limestone areas when subterranean water dissolves the limestone, leaving underground caverns whose roofs eventually collapse forming a depression known as sink hole. These holes will be filled with water and becomes small lakes. They are usually productive, because plants are able to use dissolved limestone as a source of carbon dioxide for photosynthesis.

(v) *Basin formed by streams* : Lakes may be formed by abandoned stream channels. A loop of an old meandering stream in a flat river valley is cut off when the stream cuts across in a more direct route as illustrated in Figure 1.1 (p. 9). The abandoned loop get blocked off at both ends by the deposition of sediments and becomes be a type of lake known as oxbow. Because of locations in old, fertile river valley, oxbows are among the most productive natural lakes in the world.

(vi) *Man-made lakes and ponds* : A pond may occasionally be created by digging a basin that intercept the water table. Such a pond is, in a sense, nothing but a wide shallow well. Most man-made lakes are created by damming a stream at a strategic point, so that the water backed up by the dam will be contained in a natural valley or basin.

The lakes created by flooding of river valley are usually much shallow as compared to tectonic or volcanic lakes. They are often extremely productive following their formation owing to the presence of organic and inorganic nutrients in the top soil. Since agriculture is often practiced in the watersheds, the productivity of such lakes increases rapidly with time.

1.5 ZONATION OF LAKES

Depending upon the availability of light, the lake water can be divided verticially into three major zones (Fig 1.2).

(i) *Euphotic zone* : The euphotic zone refers to the total illuminated stratum of the water and has enough light for photosynthesis. The euphotic zone extend upto the water depth at which the light intensity is 1% of the surface (called as light compensation level, where respiratory loss equalizes the photosynthetic gain) and which approximates to the lower limits of the photosynthetic activity. The light compensation level is not constant but varies with day time, season, latitude, amounts of phytoplankton and suspended materials in the water. The light compensation level is not constant for all algae, but differ from algae to algae, because different algae may have different light intensity optima for their photosynthesis, hence it is impossible to define the euphotic depth in absolute units.

The euphotic zone can further be subdivided horizontally into two subzones; (a) *Littoral subzone*, which is the euphotic zone at lake margin and is occupied principally by rooted plants and benthic algae along with phytoplankton. This zone supports a much greater number of species in many different associations and can be further subdivided into *Eulittoral* region, which is a region between high and low water marks at the water's edge, where the beating of waves is most effective and the *Sublittoral* region, which extend from the lower limit of wave action to the lower limit of rooted vegetation or euphotic depth; and (b) *Limnetic subzone*, which is the open water euphotic zone away from the shore. It contains only phytoplanktons and no benthic plants. This zone may be absent in small, shallow ponds.

(ii) *Dysphotic zone* : This is a narrow dim light zone immediately below the euphotic zone and has light intensities below 1% of the surface. In this zone the rate of respiration exceeds to that of photosynthesis and the plants will be in disadvantage if kept in this zone for long time.

(iii) *Aphotic or profundal zone* : The entire deep water zone beyond the depth of effective light penetration is called the aphotic or profundal zone. This is dark and devoid of any photosynthetic plant, hence any prolong stay of phytoplanktons in this zone may prove harmful to them. This zone has large amount of organic detritus associated with large number of aquatic bacteria and fungi on the bottom sediment, which decomposes the organic detritus and the released inorganic nutrients will be available for use by primary producers. This zone has low dissolved oxygen content.

The limnetic and profundal zones are relatively large as compared to littoral zone in lakes, but the converse is true for ponds. The profundal zone may be absent in ponds.

1.6 CLASSIFICATION OF LAKES

The lakes can be classified into different types depending either upon their (i) nutrients content and productivity, (ii) hardness of water, (iii) salinity of water, and (iv) geographical location.

(i) *Nutrients content and productivity* The lakes can be classified into four different types on the basis of their nutrients content and productivity.

(a) *Oligotrophic lakes* These lakes are usually over 18 metres depth with steep sloping sides and very little shallow water and vegetation around margin. Their waters are clear and transparent and in which the light penetrates to greater depth due to the presence of little plankton, silt or clay in suspension; and due to low nutrients content of the water, the productivity of lake water is always lower. The water of oligotrophic lakes is rich in dissolved oxygen content, low in carbon dioxide content and has alkalinity less than 40 ppm and pH less than 7.4. It has Nygaard's compound index* value of about 1 (Nygaard, 1949).

$$\text{*Nygaard compound index} = \frac{\text{Number of species of Cyanophyceae, Chlorococcales, centric diatoms, Eugelnoids}}{\text{Number of species of desmids}}$$

The absence of large amount of blue-green algae and the presence of greatest number of desmids (e.g. *Staurastrum, Staurodesmus, Cosmarium, Hyalotheca, Desmidium* etc.) as well as Cryptophyceae and Crysophyceae members appears to be an indicator of obligotrophic waters. In oligotrophic waters, the species diversity is more and phytoplankton bloom is rare. A sparse algal population will entail greater energy expenditure by predators to obtain food. The fish population is not large in oligotrophic water.

(b) *Eutrophic lakes* They are exactly opposite to that of oligotrophic lakes in all respect. These lakes are shallow with usually less than 18 metres depth with gentle sloping sides and possess an extensive littoral zone with more marsh vegetation. The water is turbid due to dense growth of phytoplankton and presence of more suspended particles, which significantly reduces the light penetration. Thus high rate of phytoplankton production is restricted only to upper waters. The colour of water may be green, yellow or brownish green.

The lake water, is rich in nutrients because of the continuous land drainage, input of sewage and phosphate containing fertilizers. Since most of the eutrophic lakes are situated in tropical regions, they have rapid rate of nutrients recycling. In eutrophic lakes, an average annual concentrations of inorganic nitrogen and phosphorus may be greater than 0.30 and 0.015 mg/L, respectively. In these lakes, the rate of gross primary production varies from 0.5 to 5 grams of dry organic materials/m^2/day during most favourable growing periods and because of the more primary production (the photosynthetic oxygen released largely escape into the air and only partly dissolves in the water) and rich deposition of dead organic matter on sediment (the decomposition requires plenty of dissolved oxygen and release carbon dioxide which may cause suffocation to or even death of fishes and other animals particularly in the hypolimnion, but a large fish population may exist in the epilimnion), such waters have high biological oxygen demand (B.O.D.).

The waters of such lakes are moderately hard and their alkalinity varies from 50 to 100 mg/L. The eutrophication reduces the algal diversity and support the massive growth of only few tolerant algae like blue-green algae *Microcystis, Lyngbya, Oscillatoria*; diatoms *Melosira, Fragilaria, Asterionella*; chlorococcales *Chlorella, Scenedesmus, Ankistrodesmus*; and dinoflagellates *Peridinium* and *Ceratium.* The contribution of desmids to the total phytoplankton biomass is very small in eutrophic lakes and the value of Nygaard's compound index is above 3. The eutrophication is a long term enrichment or ageing process of oligotrophic lakes requiring centuries but equivalent changes may also occur in decades or years due to increasing human activities.

(c) *Mesotrophic lakes* They are intermediate between oligotrophic and eutrophic lakes in nutrients content and productivity. The standing crop of vascular plants is greater in these lakes than in eutrophic lakes and the members of blue-green algae and pennate diatoms are more common than that of Chlorococcales.

(d) *Dystrophic lakes* Many lakes of northern latitudes are dystrophic. In such lakes, the water is conspicuously clean but coloured yellow to brown and may be acidic due to the presence of organic acids, humic acid and tannic acid, and may contain very low levels of carbonates. The pH of water is quite low and so is its productivity. The water has low oxygen content at all depth and contain low dissolved nutrients. The phytoplankton, benthos and fishes all are usually scarce and the lake has only few acid tolerant species of diatoms *Melosira, Eunotia, Actinella, Frustulia, Stenopterobia, Pinnularia, Tabellaria* etc. and *Chrysophytes Dinobryon, Synura, Mallomonas* etc. The dead organic matter does not decompose completely due to low dissolved oxygen content and less number of bacteria present in the water, and will undergo continuous accumulation on the bottom and will form peat which is composed of insoluble lignin, cellulose and hemicellulose. Thus the bottom of such lakes is not a true solid one but a false one.

Probably all lakes starts as oligotrophic lakes but as nutrients accumulates, the vegetation develops and the organic matter decomposes, they changes into eutrophic lakes, and if the organic matter does not decompose completely then into dystrophic lakes. The eutrophic lakes may later change into ponds and marshes and dystrophic lakes into bogs. These changes usually takes very long time usually centuries to occur.

(ii) *The water hardness*: The type of basic rock influences the water chemistry and thus on the type of lake that whether it is a soft water or a hard water lake.

(a) *Soft water lakes* : These lakes have low concentrations of total dissolved salts and relatively high content of monovalent than divalent cations. The water is poor in calcium, magnesium, nitrogen and half-bound carbon dioxide and contain not more than 5 c.c./L of fixed carbon dioxide, but often has large amounts of iron and other trace metals. The pH of water is below the neutral point. The algal flora is relatively scant and is predominantly of chlorophycean type. They are desmid habitats "par excellence". The water has fewer bacteria, zooplanktons and fishes.

(b) *Hard water lakes* : Such lakes tend to have higher concentrations of salts dominated by calcium, magnesium, bicarbonate and carbonate. They have lower content of monovalent than divalent cations. The pH of water becomes exceedingly high so that lime incrustations may occur on the submerged objects of all kinds and these lakes may have 22 to 50 c.c./L of fixed carbon dioxide. The hard water lakes inhabitat large number of blue-green algae like *Microcystis, Oscillatoria* etc. and diatoms like *Asterionella, Fragilaria* etc. The presence of *Chara, Cladophora, Spirogyra* and higher aquatic plants like *Potamogeton, Ceratophyllum, Myriophyllum* etc. also characterizes the hard or semi-hard waters.

(iii) *Salinity* : Depending upon the salinity of water, a lake may be either of saline or freshwater type.

(a) *Saline lakes* : A lake will become be saline if the solar evaporation exceeds the water input. These lakes are mostly situated at arid tropical places. They have high content of salts like sodium chloride, sodium carbonate, sodium sulphate, sodium bicarbonate etc. The water of Great salt lake, Utah, contains about 203 grams of salts per litre and the specific gravity of undiluted water is about 1.15. Saline lakes are usually alkaline and their pH can rise to more than 10 during active photosynthesis at day time. Such lakes contains only few salt tolerant algal species like green flagellate *Dunaliella salina*, some blue-green algal species of *Spirulina*, *Chroococcus* and *Anabaenopsis* and some diatoms like *Cyclotella meneghiniana*, *Coscinodiscus rudolfi*, *Anomoeoneis sphaerophora*, *Navicula* sp and *Nitzschia* sp.

(b) *Freshwater lakes* : All other lakes which are not saline are freshwater lakes.

(iv) *Latitutides* : Depending upon the latitude, a lake may be either a tropical, temperate or polar lake.

(a) *Tropical lakes* : Most of the tropical lakes are eutrophic in nature. In these lakes, the organic matter decompose and mineralize rapidly due to relatively high water temperatures. These lakes are low in species diversity represented only by few members of blue-green algae and green algae, but have about twice total algal biomass than temperate lakes. Since the climatic fluctuations are less pronounce in tropical regions, more monotonous situation with less seasonal fluctuations and fairly stable species composition occur here throughout the year. Unlike temperate lakes, which have many-fold annual variations in chlorophyll concentrations, the tropical lakes have only about two-fold annual variations in chlorophyll concentrations, with slight peak only during maximum rainfall.

The deep water tropical lakes are usually maintained in a permanent thermally stratified state because the downward mixing of the water by wind-induced turbulence is insufficient to stir the density stabilized bottom waters and the surface water may not cool sufficiently to permit complete circulation, except during the coldest period of winter. Such lakes are called as **Warm monomictic lakes**. But, in shallow water small tropical lakes, the thermal stratification, usuaully present during day time, break down at every night (due to surface water cooling of less voluminous water), and the daily repetition of such event prevent the build up of a large algal population in them and such lakes are usually called as **Polymictic lakes**.

(b) *Temperate lakes* (*subarctic lakes*) : Temperate lakes are characterized by series of events, coupled with annual climatic changes, such as: short period of ice formation on the water surface during winter, thermal stratification of water column during summer, and overturn of water during spring and autumn. Such changes imposes an immense stress on the algal population, resulting into numerous changes in species composition. Since such lakes undergo two overturns in the year, they are called *Dimictic lakes*.

An increasing levels of irradiance and the availability of sufficient silicon during spring favours the development of diatoms at that time, but their bloom gets terminated in summer due to shortage of silicon and more chances of zoolplanktons predation and fungal parasitism and are replaced by some members of green algae and blue-green algae, which in turn are replaced by some dinoflagellates. Due to disappearance of thermal stratification in autumn, which allow mixing of surface water with nutrient-rich deep water, a smaller bloom of diatoms along with some blue-green algae my develop at that time. In winter, though nutrients may be present in abundance, the decreasing levels of irradiances and low temperatures favour Cryptophycean members to grow.

(c) *Polar lakes* (*arctic and antarctic lakes*) : These lakes differ from temperate lakes in their extended period of ice cover for many months. It has been estimated that only about 20% of photosynthetically available radiation penetrated the ice and this level gets further reduced to be about 2% if the heavy snow cover the ice. Thus the snow cover on the ice results into much reduced light penetration and is followed by the upward migration of some algae, like *Synura* present in the water underlying the ice to obtain the dim light. The ice formation on the lake surface also results into a steady reduction of dissolved oxygen concentrations in the water, which is either due to low photosynthetic activity or lack of exchange with the atmosphere and continual respiratory removal. But in spite of ice cover and low light intensities, much of the carbon fixation does occur beneath the ice. The species living in ice-covered lakes are extreme 'shade' forms and can photosynthesize even at very low incident radiations. Since such algal populations are maintained in the dark for many months, the possibility of some forms of heterotrophy under these conditions can not be ruled out. Algae are partially protected from starving under ice by having a progressively decreased metabolism as the winter goes on. The grazing of algae is also extremely light during winter and perhaps also during summer. The vegetative fragments or spores of algae attached to organic matter over sediments is undoubtedly a possible way to perennate under such conditions. It is also possible that many algae may form physiological resting stages in which ice formation within the cells is prevented by accumulation of low molecular weight solutes such as polyhydric alcohols or amino acid and their derivatives.

The lakes are ice-free only for a short period in summer which has almost continuous days and about 75% of the yearly production occurs at this time. Since thermocline does not develop in these lakes and the water overturn only in the middle of summer, they are called ***Cold Monomictic lakes***. An increasing light intensities and daylength in the late spring initiated a single summer bloom of algae like *Chlorella pyrenoidosa*, *Stichococcus atomus*, *Coccomyxa coccoides*, *Chlamydomanas intermedia*, *Brachiomonas* sp., *Ankistrodesmus falcatus*, *Spirogyra* sp and *Mougeotia* sp.

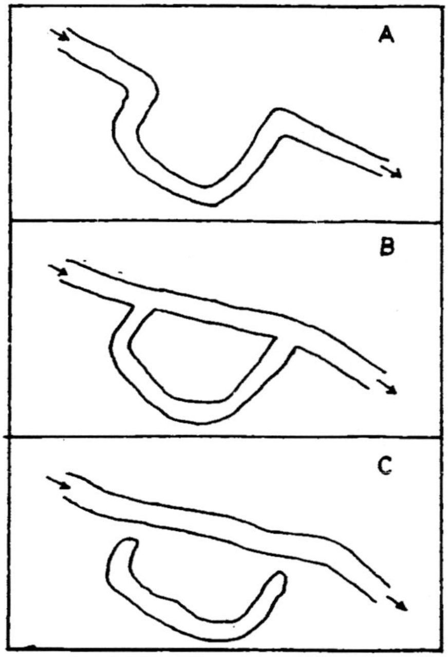

Fig. 1.1 : Steps in the formation of an oxbow lake. (A) An old meandering stream with a large loop. (B) A new channel is cut across the neck during a flood. (C) The loop is cut off from the sediment and becomes a semi-circular oxbow lake.

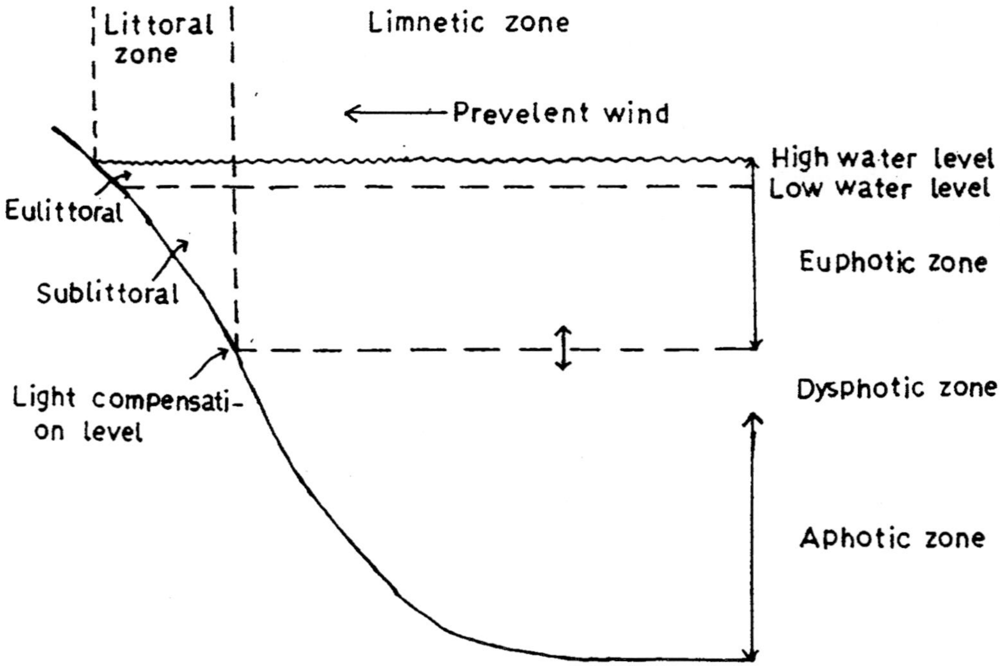

Fig. 1.2 : Major Jones in a lake.

II

LAKE FLORA: ALGAE

Other than the certain aquatic higher plants at lake margins, the algae are the principal source of organic material in the lakes. Algae make up a large part of food consumed by several species of microfauna, fishes, mollusks etc. and liberate a large amount of oxygen, vital for life support system in the lakes. About 50 to 70 percent of freshwater algae are cosmopolitan and depending upon their position in the water, that whether they are free floating or attached to a substratum, they are called as phytoplankton or phytobenthos, respectively.

2.1 PHYTOPLANKTON

The term 'plankton', coined by Hensen in 1887 and originated from Greek (literally that which wanders about) is a collective term for all organisms who floats in the water and do not execute individual movement of any importance. The phytoplankton is consisting of micro and macroscopic suspended or free floating non-motile or weekly motile unicellular, colonial or filamentous algae. The majority of phytoplanktons are non-motile and are therefore at the mercy of water turbulence within the upper water mass. However, motile phytoplanktons like *Chlamydomonas*, *Volvox*, and members of dinoflagellates and chrysophytes etc. are unable to swim against water current. The intrinsic movement of planktonic algae is usually slight relative to water turbulence induced by current. The phytoplanktons drift in a medium which is either a dilute or concentrated solution of inorganic and organic particulate, colloidal and dissolved matter derived either from drainage or from secretion, excretion and death of organisms in the water.

The phytoplanktons stand at the base line of food webs in aquatic environments and occupy the whole water mass except when the lake is stratified and a thermocline develops, when they ceases at the depth at which photosynthetically available light is 1 percent of the surface. A planktonic alga, apart from its senescent or resting cells, is usually absent from benthic situations and, in general, the converse is also true. The thick-wall sexual spores and resting vegetative cells or fragments of planktonic algae may persist for sometimes in the benthic situations and can serve as a source of inoculum for fresh planktonic growth.

Phytoplanktons are very sensitive to changes in the environmental conditions and their bloom (burst of growth) being of only few weeks duration and their species replacement takes place in a matter of month or less.

2.1.1 Classification of phytoplankton

They can be classified into different categories depending upon their size or degree of suspension in the water.

(i) Classification based upon size

Depending upon the size, they can be classified into following types (Sieburth, et.al. 1978; Sournia, 1978).

(a) *Picoplankton, Ultraplankton or μ-plankton* : They are generally consisting of solitary cells and range in linear dimension from 0.2 to 2 μm. They are present throughout the euphotic and aphotic zones and are well adapted to nutrients-depleted oligotrophic waters (in terms of more surface area to volume ratio). They also thrive, but seldom dominate, in nutrients-enriched meso-and eutrophic waters. They are examplified by *Stichococcus* sp., *Synechococcus* sp., *Cyanodictyon* sp., *Chlorella minutissima*, *Chloromonas grovei* and *Rhodomonas* sp.

(b) *Nannoplankton (Nanoplankton)*: They are mostly composed of coccoid members of Cyanophyceae and Chlorophyceae and members of Chrysophyceae and Cryptophyceae and to some extent of centric diatoms ranging in linear dimension from 2 to 20 μm. Since, they occur in such vast number that despite their minute size they have a major share in the productivity of any lake. They can not be caught completely with even the finest plankton net (but they may be retain due to clogging of pores) and as a result can only be obtained by centrifuging samples of lake water or filtering the water on fine filters such as millipore filters. They are not only abundant but also shows less seasonal variability in biomass than larger net plankton. They sink more slowly and their photosynthetic activity exceeds much more than that of net plankton (Fig. 5.7). They are ingested by filter-feeding herbivores.

(c) *Net plankton* : An algal fraction retain by a fabric of mesh aperature of 20-90 μm is referred to as net plankton and that which passes through it as nannoplankton (now generally accepted as nanoplankton). They commonly include colonial and filamentous members of blue-green algae and green algae, most of desmids and dinoflagellates and colonial chrysophytes. They are generally more concentrated near lake shore and are usually grazed by copepods. They can be further classified into three categories depending upon their size as *Microplankton*, which range in size from 20-200 μm; *Mesoplankton*, which range in size from 0.2-20 μm; and *Macroplanktons*, which are more than 2 cm in size.

Since no algal division belongs to a particular size class, the classification of phytoplanktons on the basis of size is arbitrary. For example, majority of planktonic diatoms are in the size range of 20-100 μm, but some like *Rhizosolenia* can reach upto 1-1.5 mm in length. Similarly, most dinoflagellates vary in size from 10 to 80 μm, but some like *Peridinium* and *Ceratium* can reach upto an overall size of 350 μm.

(ii) Classification based upon the degree of suspension

Depending upon that whether a phytoplankton completes its life cycle entirely suspended in the water or not, they are classified into three types.

(a) *Euplankton or Holoplankton*: Since they complete their entire life cycle suspended in the water, they are also called as true plankton, but it is very difficult to prove that they never settle down or form benthic spores. It is quite possible that their spores, which may be benthic for sometimes, when transported upward by currents or on the body of nektons (fishes etc.) act as inoculum for fresh growth. Few cells of diatom *Asterionella formosa* appears to be planktonic throughout whole year.

(b) *Meroplankton* : They are accidental planktons, which are washed into water from bottom sediments. They spend a part of their life cycle in planktonic form and the rest on the bottom sediments. Many desmids and chrysophytes form zygospores which are too heavy to be maintained in the water column.

Hutchinson (1967) pointed out that the number of planktonic species does not increase as the lake area increases, indeed a decrease may occur, and this may be an indication that many freshwater phytoplanktons are meroplanktonic, which have a more intimate association with the sediments. Conversely large shallow lakes ought to have large number of phytoplanktonic species unless shallowness imposes a constraint on the euplanktonic organisms.

(c) *Tripton, Detritiplankton or Tychoplankton*: They are those phytoplanktons which are attached to flocs of suspended particulate organic detritus of upto a millimetre or more of diameter and are transported into water by nekton. They are more common in lake water especially after a storm and they temporarily effects the habitat by decreasing the light penetration and depleting the nutrients. A euplanktonic species in a lake may be tychoplanktonic in another or even in different areas of the same lake.

2.1.2 Phytoplankton flora

Although, in general, the morphology of freshwater phytoplanktons is more or less similar to that of marine phytoplanktons, no species is common to the two habitats, except that some oceanic species may occur in some inland saline lakes. The desmids are peculiar to freshwater habitats, while blue-green algae are represented only by few species of *Trichodesmium* in seawater, but they are very diverse in freshwater habitats.

Since chemical composition of freshwater lakes vary much more than seawater, no group of algae dominates the freshwater phytoplankton. In freshwater lakes, the members of Chlorophyceae and Cyanophyceae are more abundant than those of Bacillariophyceae, Dinophyceae, Cryptophyceae or Chrysophyceae, and those of Euglenophyceae tend to be more abundant in ponds or small lakes enriched with organic matter.

Chlorophyceae : The green algae are almost invariably be found in all freshwaters. They have mostly three categories of organization i.e. unicellular, colonial and filamentous and within which exist motile and nonmotile representatives exhibiting a great range in cell size and morphology (Table 2.1 and Fig. 2.1). Green phytoplanktons exhibit a widespread distribution in terms of latitudes, but different ecotypes may occur in different geographical regions. A temperate ecotype may occur in a high mountain lake in the tropics accompanied by a decrease in temperature. Such records are further evidence of widespread dispersal of algal species across regions which are normally considered unfavourable.

In temperate lakes, the plaktonic green algae are an important component of phytoplankton at any time from late spring to autumn. Certain green flagellates are also capable to grow under ice during winter stratification, when motility is essential to maintain the cells in the euphotic zone immediately below the ice.

Since tropical lakes are subject to less seasonal variations than either temperate or high latitude lakes, in tropical lakes the chlorophytes are more or less evenly distributed in the water column and exhibit little diurnal variations in the depth distribution most of the time in the year. Green algae are characteristic component of phytoplanktons in lakes or ponds of relatively high nutrients status during periods of stable thermal stratification.

Due to their large size and biomass, desmids are present in relatively low number in water as phytoplankton, but they are usually present as plankton in regions of lakes with good water circulation. The filamentous green algae are not normally considered as true planktonic algae, but sometimes an extremely high biomass of free-floating *Spirogyra* and *Zygnema* derived from breaking off of the loosely attached benthic populations does occur. Similar to desmids, these filamentous algae are also resistant to grazing due to their large size.

Cyanophyceae: The planktonic members of blue-green algae exist either in the form of single coccoid cells or as filaments (trichomes) (Table 2.2 and Fig. 2.2). The members resembles the bacteria in lacking organized nuclei, mitochondria, chloroplast etc. and in manner of cell division by constriction. They undergo rapid rate of vegetative reproduction either by cell division, hormogonia or akinete formation and have no sexual reproduction and flagella. They have peptide glycan cell wall with usually a mucilage sheath.

Some members have the habit of floating high in the water column and can withstand extensive (daily to weekly) exposure to high (supersaturated) photosynthetically active radiation (PAR) accompanying ultraviolet radiation striking the surface they inhabit. The floating crust and scum of *Microcystis*, *Anabaena* or *Aphanizomenon* can thrive under excessively warm (35-40°C) and high PAR conditions in the surface waters and can prevent the passage of light into water and can adversely affect the photosynthetic production of planktonic algae and thus the dissolved oxygen content of the water. When floating crust of algae die and disintegrate after a few hot days and nights in the summer, they will further aggravate the dissolved oxygen problem in the water and will adversely affect the survival of fishes. However, some members like *Oscillatoria*, *Lyngbya*, *Phormidium* and *Synechococcus* often inhabit distinct subsurface layers, but can occasionally alter their vertical position in the water column by buoyancy regulation through gas vacuoles to seek optimum light and nutrients regimes suitable for their growth.

Blue-green algae are often sole inhabitants of extremely nitrogen-deficient waters, because of the ability of certain members to biologically fix atmospheric nitrogen into ammonium by means of an oxygen sensitive enzyme complex nitrogenase. In heterocysts, he elimination of photosystem II activity negates photolysis of water, which normally lead o molecular oxygen production. The potential oxygen inactivation of heterocyst nitrogenase activity is thereby eliminated. The heterocysts undergo high rates of endogenous respiration, ATP synthesis and hydrogenase activity and such enhanced metabolic activities appears o be directed at producing energy required for nitrogenase-mediated nitrogen fixation as vell as scavenging and reduction of molecular oxygen, which can potentially inhibit nitrogenase activity. Despite the protection provided by thick cell envelopes and polar plugs, hese morphological adaptations are not absolutely effective in shielding oxygen.

Several non-heterocystous blue-green algal members are also capable of nitrogen fixation. These organisms require oxygen-free conditions near the site of nitrogen fixation. The filamentous genus *Oscillatoria* appears to compensate for the absence of heterocysts either temporally by confining most of their nitrogen-fixing efforts to nonphotosynthetic periods (night time) or spatially by growing in colonies which are physiologically differentiated or are closely associated with bacteria which are capable of locally removing oxygen around filaments via respiration. The localized oxygen consumption by bacterial epiphytes near oxygen-sensitive heterocysts or nitrogen-fixing loci in non-heterocystous blue-green algae can help protect nitrogenase from oxygen inactivation, consequently promoting optimal rates of nitrogen fixation in nitrogen depleted, oxygenated waters.

Several single-celled and filamentous blue-green algae reside either in dimly-lit hypolimnion or in periphyton communities, where their metabolic and energetic needs for reduced carbon compounds may be met either by extremely low rates of photosynthesis (accompanied by low oxygen evolution and growth rates) or by heterotrophic utilization of ambient soluble organic compounds. In this manner, low but uninhibited rates of nitrogen fixation could be maintained. Diverse blue-green algae readily assimilate a variety of organic compounds at naturally occurring trace quantities. *Anabaena oscillariodes, A. circinalis* and *Microcystis aeruginosa* are capable of assimilating glucose, maltose, galactose and variety of amino acids occurring at trace (natural) concentrations.

Several members are able to store phosphorus as polymeric polyphosphate bodies, when ambient orthophosphate supplies exceeds the growth requirement, and thus can survive and maintain growth during external phosphorus deficiency.

Many members also reveal a high degree of tolerance or a favourable growth response to excessive nutrients or pollutants loading. A large number of blue-green algae shows a negative growth response to acidic conditions and have a distinct preference for neutral to alkaline conditions and are rapidly replaced by chrysophytes and chlorophytes under acidic conditions.

Some blue-green algae like *Anabaena flos-aquae, Microcystis aeruginosa, Aphanizomenon flos-aquae* and *Nodularia spumigena* produce metabolites or decay products which are toxic and can cause painful gastrointestinal upset, vomiting, fever or muscular pain in animals who drink water from ponds or lakes containing their dense algal bloom.

Bacillariophyceae : Some of the common freshwater planktonic diatoms are *Asterionella, Fragilaria, Tabellaria, Diatoma, Cyclotella, Melosira* etc. (Table 2.3 and Fig. 2.3). They exist in the form of single cells, chain of cells or stellate or other alignment of cells. Their cell wall, consisting of two overlapping pieces, is composed of amorphous hydrated silica plus organic components. Due to slower recycling of silicon, the rapid growth of diatoms will quickly reduce the Si: P ratio in favour of other taxa, unless deep mixing bring some silicon from deeper water to surface water. Thus sinking would be of greater evolutionary importance to diatoms than to other planktonic algae. It facilitates nutrients uptake by reducing nutrients–depleted microzones and also act as an escape mechanism against predation. Most of the diatoms prefer eutrophic waters. The planktonic members may have large central vacuoles which may aid to flotation. The blooms of diatoms in temperate waters during spring and autumn seasons are mainly due to availability of more silicon at that time due to water overturn. They persist unfavourable periods by maintaining a slowly declining stock of resting vegetative cells either suspended in water or lying on the bottom sediment.

Dinophyceae : The most common freshwater members are *Gymnodinium* spp, *Peridinium cinctum, P. pusillum, P. willei* and *Ceratium hirundinella* (Fig. 2.4). The cells of *Gymnodinium* are broadly oval and have no wall and several of its species are known to produce a lethal toxin which may cause death of fishes. The cells of *Peridinium* and *Ceratium* are representative of armoured dinoflagellates and are covered with cells walls of closely fitting cellulosic plates. The cells of *Peridinium* are top-shaped, while that of *Ceratium* are distinct because of their prominent straight or curved horns. *C. hirundinella* has one long anterior horn and three (or two) posterior horns, all of different lengths. The dinoflagellates are relatively large organisms (*Peridinium*, diameter 30-70 µm; *Ceratium hirundinella*, length 90-450 µm) and have, therefore, a low surface area to volume ratio and can not compete effectively with small algae as far as nutrients uptake is concerned, but they are motile and shows a circadian vertical movement and are thus able to absorb nutrients from different depth. The cells of dinoflagellates have a pair of unequal heterodynamic flagella, in which the larger one emerge from a groove called sulcus and pushes the cell forward from behind and the shorter one which is a ribbon shape, lie in a transverse groove called girdle and provide a rotating motion.

The cells have a characteristically large nucleus, containing high number of permanently condensed chromosomes (may be upto 500) and numerous golden-brown oval shape chloroplasts. The cells reproduce by simple fission, as do diatoms, but in dinoflagellates each of the daughter cells retain half of the original cellulose armour and forms a new half to replace the missing half without any diminution in size. The dinoflagellates prefer eutrophic waters and produce bloom in temperate lakes in summer season. The period from March to July is characterized by their exponential growth, followed by a stationary phase during July to September and a rapid decline in population occurring in late September or early October. The population persist as benthic cyst during winter in temperate lakes.

Chrysophyceae : The members of chrysophyceae are unicellular or colonial, bi-or uniflagellate with anterior flagellum possessing complex lateral hairs. The cells have two to many chloroplasts and an eye spot lying outside the chloroplast. The cells are often naked with only cell membrane (e.g. *Chromulina, Ochromonas* and *Uroglena*) or are covered with silicified scales and bristles (e.g. *Mallomonas, Synura* and *Chrysosphaerella*) or by an organic lorica which is sometimes secondarly mineralized (e.g. *Chrysococcus* and *Dinobryon*) (Table 2.4, Fig. 2.5). The scale-covered and loricate chrysomonads swim in the water more slowly than the naked species of equivalent cell size.

A number of chrysophytes like *Mallomonas, Dinobryon, Chrysococcus, Synura, Ochromonas* etc. undergo diurnal movement over a vertical distance of 0.5 to 2 metres. They migrate upward in the morning and downward around noon or late afternoon depending upon the sunny or cloudy days, respectively. *Uroglena* is perhaps the strongest swimmer among chrysophytes and is the only form capable of overcoming epilimnetic turbulence to any extent. The ability to overcome ambient turbulent mixing and to move directionally in response to nutrients or light gradients has been proposed to be an important competitive advantage to flagellate algae in some habitats. Small chrysomonads did not show any evidence of nutrients limitation because of their high surface area to volume ratio and small biomass which makes them a superior nutrients competitors.

Chrysophytes prefer low light and low temperatures and their predominance year-round in arctic and north temperate lakes or in cool summer metalimnion seems to support such a correlation. Chrysophytes are codominant with other flagellates in softwater oligotrophic and mesotrophic unpolluted circumneutral arctic or alpine lakes or ponds, but have also been recently reported to be an important planktons of tropical oligotrophic lakes.

In highly eutrophic situations, where phosphorus is always in excess, chrysophytes would certainly be outcompeted by green algae. Chrysophytes dominance in oligotrophic lakes may very well be dependent on the lack of large efficient cladocera grazers such as *Daphnia*.

Chrysophytes are nutritionally opportunistic, being able to switch between autotrophic, heterotrophic and phagotrophic modes in response to cells requirement and environmental conditions. Many photosynthetic species have an obligate requirement for several B vitamins and can supplement their photosynthetic carbon fixation with the assimilation of simple organic molecules such as glucose, fructose, sucrose, pyruvate and glycerol. Almost every genus of planktonic chrysophytes tested, including those which are photosynthetic or colourless (except *Mallomonas* and *Synura*) are capable of actively ingesting the treated beads, bacterial cells or other algae. It is an intriguing possibility that leakage of organic molecules from Chrysophytes in combination with Chrysophyte bactivory may represent a nutritional strategy that makes these algae particularly adapted for persistence in low-light habitats.

Several species of Chrysophytes produce an extracellular organophosphate degrading enzyme (acid phosphatase) in response to phosphorus limitation or in presence of organophosphate under acidic condition, but they do not secrete significant quantities of alkaline phosphatase in response to P-limitation under alkaline conditions as is typical of many other microalgae, and perhaps the lack of this ecologically important enzyme contributes chrysophytes observed preference for neutral or slightly acidic waters. The sensitivity of elevated pH could be related to their inability to secrete alkaline phosphatase enzyme and also to utilize bicarbonate as a source of carbon for photosynthesis. They have perhaps an obligate requirement for dissolved free carbon dioxide as the only form of dissolved carbon used in photosynthesis. The near absence of chrysophytes from high-alkalinity, hard water lakes might be explain by presence of low levels of dissolved organic carbon in such lakes. Conversely, the great importance of chrysophytes in highly coloured lakes rich in humic and fulvic acids may also be related to the presence of readily available forms of iron or other micrometals in them. Chrysophytes reproduce mainly through binary fission, but under some conditions individual cells can be induced to form a statospore.

Cryptophyceae: Although the cryptomonads are found in lakes of antarctic mainland and in temperate and tropical regions, they display the largest diversity in spatial and temporal niche occupation in lakes of temperate regions. They are not recorded in hypersaline lakes or in hot springs. They have their maximal population development well below the water surface between 15 and 23 metres depth. Thus they are capable of utilizing low light levels, but are less able to deal with high pH. Their bimodal (spring-autumn) population maxima in temperate lakes may either be due to water fortified with nutrients from deeper strata or from temporary relief from predation, when populations are diluted into larger volumes of water due to water overturn. The under-ice survival in winter seems to be due to low respiration rates and low grazing pressure at that time.

They are free-living, unicellular, asymmetrical dorsiventrally flat (the dorsal margin is convex and the ventral margin is usually flat or concave) highly motile flagellates. Their cells are bound by a firm but elastic periplast, which is a plasmalemma with granular or fibrillar material deposited both to its outside and inside. The cells have two somewhat unequal flagella arising from their anterior end either from a furrow or a gullet. In most genera, e.g. *Cryptomonas, Rhodomonas* etc., there is a tubular gullet extending from the anterior end of the furrow, but in some like *Chroomonas*, there is no gullet and only a furrow is present. The cells have a nucleus situated near the posterior extremity, and have either one (*Rhodomonas*) or to (*Chroomonas, Cryptomonas*) chloroplasts and rounded pyrenoid-like bodies, apparently not embedded in chromatophores. A prominent contractile vacuole discharging its content either into furrow or gullet is present (Fig. 2.4.). The trichocysts are present near cell periphery and depression. The transparency of the living cytoplasm allow all intracellular organelles to be recognized. The cells have an obligate requirement for one or more vitamins and reproduce through oblique binary fission. They impart a foul odour to the water and sometime produce fish-kills due to some secretory products.

They are among the most heavily grazed plankton by calancid *Eudiaptomus* and are generally present in the gut content of *Daphnia hyalina* and *Diaptomus gracilis*. A few of them may embed themselves in palmelloid mucilage layers or form resting spores to avoid grazing animals. They may also have enemies of parasitic nature. A phycomycete *Rhizophydium fugax* may infect the extracellular polysaccharide of *Cryptomonas* and can harm the cells.

Euglenophyceae : The members of euglenophyceae are commonly present in ponds with high organic content particularly in summer and autumn seasons. Some of common members are *Euglena, Trachelomonas* and *Phacus* (Fig. 2.4). They are bi-or uniflagellate cells, and the flagella (with hairs) originate from an apically situated depression called gullet. The cells are covered by intracellular covering consisting of spirally arranged protein strips. A red eye-spot lie free in the cytoplasm and is not embedded in the chloroplast which may be one to many in number. The products of assimilation appears as solid, often large granules of paramylon of very diverse shape. The members perform characteristic euglenoid movements. Some non-photosynthetic members are also common.

2.1.3 Spatial distribution of phytoplankton

Phytoplankton are rarely found uniformly distributed in any water body. A water sample from the surface may contain a different association of species than one from at 1, 5, 10 or 60 metres depth. A characteristic feature of phytoplankton is that they tend to vary spatially on a scale of less than one millimetre to ten to even hundreds of kilometres both horizontally and vertically as well as daily, seasonally and annually. These patches are generated by changes in physical, chemical or biological factors.

(i) Horizontal distribution

The horizontal distribution pattern of phytoplankton in any lake is influenced by many factors like shape and size of lake, extent and direction of water currents, position of inflowing streams, irregularity of shoreline and swarming of a particular species in local area. But the dominant factor influencing the gross heterogeneity in horizontal distribution is the circulation pattern of water induced by wind. A light northern wind will transport the surface phytoplanktons to the southern end of a lake within few hours. The active swimming by certain zooplanktons can also cause phytoplanktons to aggregate in dense groups.

The horizontal variation in phytoplankton composition is usually slight in small lakes, but it may be prominent in very large lakes. There are mosaic of communities moving with water currents. The greatest difference may be in the shallow littoral zone where there is a possibility of contamination from benthic flora and in the inflow region where the new water may contain species washed in from the inflow.

(ii) Vertical distribution

The composition and abundance of phytoplankton vary more striking in a vertical direction than in horizontal direction. The factors contributing to the vertical variations or patchiness are mainly the changes in light intensities, temperatures and nutrients content with increasing depth. The rate of phytoplankton cell division is found to be comparatively more in sub-surface waters having low light intensities than in surface water having high light intensities.

The vertical variation in phytoplankton distribution is more striking during thermal stratification of water. The effective depth to which a phytoplankton can grow in a stratified lake is reduced to the depth of epilimnion. Although, the water currents continually circulate the phytoplanktons in the epilimnion, there may be a vertical variation within the epilimnion itself. Those phytoplanktons which are present below the epilimnion are actually trapped in a zone where water current slow down and the rate of algal sedimentation is increased and only after the breakdown of thermal stratification, they are distributed relatively evenly throughout the water column. It is also likely that differential sinking rates or buoyancy mechanisms of different phytoplanktons may play a part in maintaining vertical variation in phytoplankton distribution.

In some ponds and small lakes, hydrogen sulphide rich layers occur at the sediments and where only some blue-green algae and euglenoid persist and grow. The ability of zooplanktons to migrate vertically and to graze phytoplanktons at a rapid rate certainly creates their patches.

2.1.4 Distribution and species composition of phytoplanktons with change in time

The horizontal and vertical distribution of different phytoplanktons may undergo changes with change in time. The temporal changes in boundaries and species composition of phytoplankton assemblage are affected, beside by the scale of measurement, also by the time of day, seasons or yearwise.

(i) Diurnal Fluctuation

The nutrients uptake, the photosynthetic ability, the cell division and the vertical migration of phytoplanktons, all are entrained by light-dark cycle.

(a) *Nutrients uptake, Carbon fixation and cell division* : Although, each algal species has its own characteristic time course for nutrients uptake and cell division, in most cases the uptake of nutrients is light dependent (the nutrients uptake rate increases before sun rise and declines before night), and the dark period or night time is the time at which cell elongates, chromatophores divides and cell divides. In general, a peak of carbon fixation occurs between mid morning and early afternoon, followed by a depression in late afternoon due to internal end-product inhibition (Fig 2.6). The diurnal variation in carbon fixation is certainly quite marked at equator than towards northward or southward.

(b) *Vertical migration* : The possession of flagella and the ability to move is advantageous to phytoplanktons under calm water conditions or under conditions of very restricted water turbulence (e.g. during thermal stratification), and this may be expressed by their phototactic response maintaining them within a suitable light regime and also enabling them to obtain nutrients from different depth strata in a lake. The ability of phytoflagellates to swim enable them to overcome the loss of actively growing population by sinking. However, the advantage of motility may be easily outweighed by increased water movements due to wind stress, but the nonmotile phytoplanktons exhibit a greater net rate of population increase when present in turbulent water, because water turbulence maintain them suspended in the water and prevent their sinking.

In the nature, dinoflagellates are concentrated in the upper-most 5 metres of water and they migrate towards water surface during day time, often collecting around 2 metres depth and descend or disperse at night. The time taken to traverse a particular distance varies from species to species. The circadian vertical movement results mainly from the active swimming and is correlated with the light irradiance and the chemical composition of the water during quiescent periods. *Peridinium* cells performs an active upward movement from late afternoon until midnight. However, intense wind may push them downward and they may aggregate at thermocline. The vertical migration rhythm is not as well documented in green flagellates as it is in other algal flagellates like dinoflagellates.

An algal cell which is stationary in the water column soon depletes nutrients in the microlayer near the cell surface until it is replenished slowly by diffusion of nutrients from the surrounding water, but as the cell sink or rise in the water column creating turbulence and renewing the water next to the cell, microgradients are broken down and the cell surface is exposed to a greater volume of freshwater resulting into more nutrients uptake. Thus vertical migration would give such algae a competitive advantage.

(ii) Seasonal fluctuation over a year

The seasonal fluctuations in phytoplankton abundance and species composition in any water body is due to differential response of different species to changing levels of light, temperature, nutrients, grazing pressure, onset of parasitic infection, extracellular metabolites of plants and animals with a change in season during a course of year. The seasonal fluctuations in number and kind of phytoplanktons are more pronounce in temperate or polar lakes than in tropical lakes.

(a) *In temperate lakes* : Striking changes in phytoplankton abundance and species composition occur in such lakes along with a change in season (Fig. 2.7 b).

Winter season : phytoplanktons mostly consisting of some members of diatoms, desmids and cryptophytes are present sparcely. Their productivities are lower due to low temperatures and light intensities.

Spring season : An increasing irradiance levels (higher light intensities and longer days) in the early spring lead the development of phytoplankton bloom dominated by diatoms, until it disappears due to depletion of nutrients (particularly silicon), predation by zooplanktons and the conditioning of the medium by preceding organisms, imposed by thermal stratification in the summer. Chrysophycean flagellates carrying silica scales often follow diatoms bloom in annual succession of planktonic algae, because they are adapted to low levels of silicon than the diatoms.

Summer Season : Due to increasing light intensities and temperatures in the summer, the diatoms are replaced by green algae, dinoflagellates and blue-green algae. The summer minima of these phytoplanktons are mainly due to their intense grazing and fungal parasitism, since they are circulating in a smaller volume of water above the thermocline and are close to one another.

Autumn overturn : As the surface water cool down and the thermocline disappears, allowing the mixing of nutrients-poor surface water with nutrients-rich deeper water, a second smaller bloom of diatoms and blue-green algae may appear in autumn. The fall in algal population in late autumn may be due to availability of reduced light intensities, greater mixing of water which takes algae intermittently below the photic zone and increased grazing pressure.

(b) *In polar lakes* : Arctic and Antarctic lakes are surface frozen most of the time in the year and the ice resulting from low temperatures play an important role in decreasing light penetration and producing mechanical injury and thus exerting a great selectivity of species. The arctic and antarctic algae are partially protected from starving under ice by having a progressively decreased metabolism as winter goes on. A single peak of phytoplankton growth in summer is initiated in late spring due to increasing levels of light intensities, daylength and temperatures (Fig. 2.7 a). Thermocline usually does not develop here and nutrients are not limiting at any time. In northern Swedish lakes, chrysophytes like *Chromulina, Mallomonas* and *Dinobryon*; cryptophytes like *Rhodomonas* and *Cryptomonas*; dinoflagellates like *Gymnodinium* and *Peridinium*; chlorococcales like *Sphaerocystis, Crucigenia, Oocystis, Elakatothrix, Planetococcus, Ankistrodesmus*, *Chlorella* and *Stichococcus* and typical diatoms and desmids flora prevail and grow during summer season.

(c) *In tropical lakes* : In contrast to temperate and polar lakes, the tropical lakes do not experience marked seasonal fluctuations and thus do not exhibit variations in stocks of phytoplankton species composition. The phytoplankton abundance is regulated by fairly equal influences of any factor like nutrients supply, water turbulence, grazing pressure etc. throughout year. The tropical lakes, in general, support a higher biomass of algae, but shows less species diversity than temperate lakes (Fig. 2.7 c). The tropical lakes which are eutrophic in nature have continuous blooms of blue-green algae and green algae throughout year, but diatoms are relatively rare.

(iii) Long term fluctuation over many years

Although annual succession of phytoplankton is usually very consistent with the same species appearing year after year in a particular lake; in certain ditches and lakes, various algal species appear, disappear and reappear in an erratic manner year after year, and the succession pattern of phytoplankton species changes nearly every year. For example, phytoplankton succession pattern in lake PluBsee has been found to be variable year after year (Hickel, 1978; 1985). In summer 1969, *Anabaena flos-aquae* dominated. In summer 1974, a bloom of *Oscillatoria redeckei* developed and was followed by *Cryptomonas, Ceratium* and *Oscillatoria* again. In summer 1976, *Oscillatoria* and *Peridinium elpatiewsky* developed and in summer of 1981 and 1982, water blooms of *Ceratium furcoides* and *C. hirundinella* developed. Such variations are probably correlated with year to year changes in the environmental conditions as well with in the dispersal and distribution of algae.

2.1.5 Phytoplankton circulation in water

In temperate lakes, during autumn and spring seasons, thermal stratification break down and the whole water is subjected to overturn, which not only distribute nutrients homogeneously in the water but is also responsible to carry phytoplanktons deep into water. The depth upto which a phytoplankton cell circulate is called the mixed depth. Since, mixed depth is often two to five times greater than the euphotic depth, a phytoplankton cell is forced to spend sometimes in semi-darkness or total darkness (Fig 2.8). As long as the time spent by phytoplankton in semi-darkness or total darkness is not excessive, the gross photosynthesis of vertically circulating phytoplanktons will be in excess to that of total respiration and there will be some net gain, but if this time exceeds a certain level (which will certainly depends upon the depth of mixed depth), the phytoplankton respiratory losses may equalize the gross photosynthesis, resulting into no net gain and the mixed depth at which this occurs is called the critical depth. Thus critical depth is the depth of water at which the gross photosynthesis of vertically circulating phytoplanktons will equalize their total respiration. It will naturally be much deeper than euphotic depth (the depth at which no gain will be observed, if phytoplanktons are kept constantly there for all time). Both euphotic depth and critical depth lie at deeper points during spring and summer seasons (due to intense light intensities) than in autumn or winter.

Whenever the wind-driven vertical mixing of phytoplanktons is less than the critical depth, the photosynthesis will exceed the respiration, but if it is more than the critical depth or in other words the phytoplanktons spend more time in darkness, the respiration losses will exceed the gross photosynthesis and the algae will be adversely affected. Thus deeper a lake will be, the greater will be the time spent by a phytoplankton in darkness and lesser will be the gross production. The possible reason of slow growth of phytoplanktons during spring may be due to their deep circulation at that time prior to establishment of thermal stratification, but as thermocline develops during summer, the retention of the phytoplankton circulation in the epilimnion probably results the cells remaining in an advantageous position as far as the light is concerned, but due to comparatively smaller volume of epilimnion they are subjected to decreased nutrients availability and increased grazing pressure and fungal parasitism and their population growth is again checked to a greater extent.

2.1.6 Phytoplankton bloom

Formation

The increase in number or aggregation of masses of cells, filaments or colonies of planktonic algae in surface water is involved in bloom formation. The blooms of blue-green algae (e.g. *Microcystis, Anabaena* and *Aphanizomenon*) favours mildly alkaline, warmer and slightly eutrophic waters and a period of calm weather, while that of green algae and euglenoids favours excessively eutrophic and polluted waters. In temperate lakes, diatoms undergo bloom formation in spring and autumn seasons; and in polar lakes, chrysophytes, cryptophytes, chlorococcales and desmids develops bloom formation in summer season.

Disappearance

The algal bloom generally disappears rapidly due to photosynthetic inhibition and increased senescence and death of cells caused by different physical, chemical and biological factors such as intense sunlight, high temperatures, intense wind, nutrients deficiency and crowding of cells. But surface bloom of *Microcystis* develops high levels of carotenoids which may protect the alga against dangerous levels of ultraviolet light.

Influences

(i) On dissolved oxygen content

The floating mats of algae on the water surface may act as a barrier to the diffusion of atmospheric oxygen into the water, and the oxygen produced by them largely escape into the air. However, those algal bloom which are dispersed in the water column will generally increase the dissolved oxygen content of the water. Since nitrogen fixation by blue-green algae seems to decrease as the oxygen content increases, such dispersed algal bloom may be detrimental to nitrogen fixation metabolism. The oxygen helps to prevent foul or septic conditions by favouring the activities of aerobic rather than anaerobic bacteria, but when the algal bloom dies, the dissolved oxygen content of the water is used up in the decomposition of organic matter.

The dense algal bloom on the water surface causes shading in the water column and can consequently reduces the carbon fixation and oxygen liberation of the submerged plants and can produce their death. It is also not uncommon to find many moribund cells on the under surface of floating water blooms due to self-shading produced.

(ii) Nutrients concentration

The aggregation of phytoplanktons in close proximity can be harmful, since nutrients are more easily depleted in waters having algal bloom.

(iii) Toxin production

Several types of algal extracellular products including alkaloids and cyclic polypeptides have been identified in the water having bloom forming species. In general, the blue-green algae are more common cause of fish mortalities in freshwater than dinoflagellates, but the intense bloom of dinoflagellate *Ceratium hirundinella* can also cause fish-kill.

(iv) Bacterial and zooplankton populations

During phytoplankton blooms, the zooplanktons may exhibit superfluous feeding and the food they ingest is only partially digested before being voided and this rich faecal material is rapidly recycled by bacteria. Thus bacterial numbers, often, though not invariably, rises during and after algal bloom, because of either direct effect of exudation by algae or by secondary faecal effect. The bacteria produces the carbon dioxide, which might be utilize by algae and it is more likely that this relationship is brought about by oxygen consumption by bacteria. The zooplanktons usually respond to phytoplanktons blooming by increasing their own number, since they largely depend on the latter, but their population growth lag behind the phytoplankton growth to some extent because of the time required by their adult to mature.

2.1.7 Phytoplankton flotation

The specific gravity of most freshwater is 1.0027, but that of algal cells protoplasm is 1.01-1.05 and it is further increased by silicified or cellulosic walls. The silica wall of diatoms has a density of 2.6, whilst the cellulosic wall of the dinoflagellates and desmids has a density of about 1.5. Thus the paradox is that that most phytoplankton organisms are too heavy and there must be some compensation for this overweight if they are to remain in suspension. The phytoplanktons must maintain themselves in the water column where irradiances will support photosynthesis sufficient for net growth and reproduction. They must also have some tendency to selective sinking which may promote nutrients uptake and avoid predation and high levels of irradiances at the water surface. Some of the mechanisms which may help phytoplanktons to float are given below.

(i) Intrinsic properties of living cells

It has been observed that cells killed artificially have a greater rate of sinking than live cells of the same species. Thus there may not be any apparent flotation mechanism in the cell, but some intrinsic properties of living cells may help them to float and if this would not have been so, the number of senescent and dead cells in the epilimnion would soon outnumber the live cells. The sinking rate is more in older cells than in young cells.

(ii) Water turbulence

No matter now calm a water mass appears to be, it is likely that some small scale turbulence is taking place all the time. Winds of varying forces and directions are persistent feature, and even when minimal there is still a significant effect of wind on the water. The water turbulence produced either by wind or periodic diurnal temperature variations between surface and sub-surface waters help phytoplanktons to be near the water surface where photosynthesis is most effective. Water turbulence dilute the algal extracellular products and maintain a constant supply of nutrients to algae. In the absence of water turbulence, most of the non-motile phytoplanktons will sink in the water. When ice forms on the water surface, most of the heavy forms like *Melosira* will sink. The presence of non-motile phytoplanktons like *Asterionella* in the surface water is related to water turbulence. The water turbulence shows an inverse relationship with depth.

(iii) Viscuous drag

The viscosity of water is related to its temperature and salinity. As the temperature of water decreases and its salinity increases, the viscosity of water increases and with that an organism will sink less rapidly in it. Hence, theoretically, sinking of an organism increases during spring and summer and is at its lowest rate in the mid winter. A decrease rate of sinking has been observed in rich media or in saline lakes.

(iv) Reduction of overweight

Phytoplanktons can reduce their overweight by any of the following means.

(a) *Gas vacuoles* : The surface blooms of blue-green algae are a well known feature of many lakes and ponds during calm conditions. The flotation devices in the blue-green algae occurs in the form of gas-filled vacuoles which accurately regulate their position in the water column at the depth most favourable for their growth, but it must not be assumed that all vacuolate blue-green algae floats to the water surface. There are some blue-green algae which are not planktonic but have gas vacuoles. The other functions suggested for gas vacuoles are light shielding properties, storage of metabolic by-products and bubbles of gases produced.

The cells having experienced poor photosynthetic histories often tend to exhibit maximum buoyancy, whereas optimal photosynthetic histories lead to negatively buoyant cells. Accordingly, cells having relatively low rates of photosynthesis tend to increase buoyancy because gas vacuoles formation can proceed in a relatively unhindered fashion and cell ballast accumulation would be minimal. In contrast, cells having recently maintained optimal photosynthetic rates contain relatively large amounts of ballast and few gas vacuoles.

The mechanism by which cells adjust their position in the water column is assumed to be by adjusting the volume of gas vacuoles to that of cell. Algae like *Microcystis*, *Anabaena* and *Anabaenopsis* concentrate in the surface water in the morning and sink to bottom every evening. Such rapid changes in buoyancy is accounted for by the collapse of gas vacuoles due to rise of turgor pressure to several atmosphere, brought about by accumulation of photosynthate and uptake of potassium ions during intensive photosynthesis at day time, and their regeneration by self-assembly process in dim light at the bottom of lake. The gaseous content of the gas vacuoles is in equilibrium with the surrounding water. However, *Oscillatoria*, *Phormidium* and *Chrococcus* are commonly found in definite sub-surface metalimnetic layer or in mid water position and it is presumably due to subtle combination of gas vacuoles formation and discharge, which achieves a balance in such a low light intensity zone where photosynthesis proceeds at a reduced rate.

(b) *Oil and fat* : It is generally assumed that fat is an agent of buoyancy presumably since oil floats on the water and certainly an alga such as *Botryococcus* does float in still water. The fat accumulation in phytoplankton cells is regarded as a mean of counter-balancing the excess density produced by mineralized cells walls. The fat ladden cells of freshwater diatom *Nitzschia palea* are found to sink at an appreciably slower rate than normal cells lacking extensive fat deposition, but the efficiency of oil droplets in diatoms as flotation device has been questioned, since excess weight of silica walls of diatoms probably outweigh the buoyancy effect of small amount of oil occurring in the cell. The oil and lipid also accumulate in benthic diatoms suggesting that their aid in reduction of sinking may be incidental.

(c) *Ionic adjustment of vacuolar sap* : Vacuoles are usually well-developed in planktonic diatoms and may serve to reduce the average density of cells. The selective accumulation of light monovalent ions, e.g. Na^+, NH^+_4 etc. and the elimination of heavy divalent ions, e.g. sulphates etc. from the vacuolar sap of diatoms may reduce the specific gravity of vacuolar sap, but the ionic regulation of vacuolar sap is an unlikely mechanism for freshwater plankton flotation, because of the low salt content of the fresh-waters.

(d) *Mucilage sheath* : The possession of mucilage sheath by many large colonial nonmotile chlorophytes and by some desmids may confer their lower specific gravity and lower sinking speed in the water column.

(v) Size of organism

Phytoplanktons can reduce their sinking rate by increasing the surface of resistance against water. We know that surface area increases as the square of the linear dimension, whereas volume increases as the cube. This means that smaller an organism is, the greater will be its surface area relative to volume. Thus, by remaining small, planktons offer far more surface area of resistance to sinking per unit volume of living material than if they were large. This is particularly important for tropical phytoplanktons, because they live in warmer and hence less viscous water, where sinking would be faster. In diatoms, the larger cells are presumably the postauxospore cells and one might expect, therefore, a greater loss of cells by sedimentation immediately after auxospore formation and a massive bloom of only moderate or small size cells.

(vi) Shape of organisms

If the volume and mass of a cell remain constant, any morphological change which increases its surface area, will also increase its friction against water.

(a) *Superficial outgrowth*: A variety of superficial outgrowth and protuberances are borne by phytoplankton cells. They provide a high surface area to volume ratio and will increase the frictional drag of organisms against water, but if they are silicified or calcified, their excess density would counteract any buoyancy advantage gained through increased frictional drag. The delicate spines or spine like shape of entire cells help to keep certain non-swimming green algae such as *Actinastrum*, *Micractinium* and *Scenedesmus* suspended in the water. Since, the formation of spines by *Scenedesmus* depends upon the concentration of nutrients media, the alga may possess differing capacities for flotation with the change in season.

(b) *Colony formation* : The colony formation usually considered as an adaptation to increase buoyancy, actually results into increased sinking rates because of the decrease in surface area to volume ratio. However, complex shapes and ornamentation of colonies can increase the surface area to volume ratio and can produce a slightly decreased sedimentation rate in comparison to a sphere of similar cell volume. Many planktonic algae can form complex colonies, e.g. diatoms *Asterionella, Fragilaria* and *Tabellaria* and green alga *Pediastrum*, and it is very difficult to calculate their form resistance or flotation adaptation under varying conditions in nature.

2.1.8 Osmotic regulation in phytoplankton

Freshwater organisms have a definite problem to solve with regard to osmoregulation. In general, freshwater algae have an osmotic potential equivalent to about 0.15 M NaCl, significantly above that of most freshwater environments. Since, the concentration of salts is greater in the internal fluids of cells than in the freshwater environment, the water either tends to enter the cells by osmosis or solutes rapidly leak out of the cells, before water influx burst the cell membrane. To control the osmotic pressure, contractiole vacuoles are present in many thin-walled freshwater flagellates like *Chlamydomonas*, euglenoids etc., while paired vacuoles termed pusules are thought to be excretory or osmoregulatory organelles in dinoflagellates. However, cells with strong cell wall use turgor pressure to counteract any net water influx.

Phytoplankton cells adapt the osmotic stress of changing salinities by adjusting their internal solute concentrations, in order to return to equilibrium with the environment. When cells are exposed to a sudden increase in salinities or ionic concentrations of the medium, they synthesize variety of organic solutes such as glycerol (in euryhaline *Dunaliella*), galactosyl glycerides (in chrysomonad *Ochromonas*), mannitol (in *Platymonas*) and proline in diatom *Cyclotella meneghiniana* in order to return to equilibrium with the environment.

2.2 PHYTOBENTHOS

They are those algae which live on the surface of a substratum of some sort and are not free-floating in the water. The extent of benthic environment for algal growth comprises the zone of a lake shore which is sufficiently illuminated for photosynthetic purpose and may extend from few centimetres above the mean water level to either only a few centimetres below the mean water level in turbid waters or to several metres depth in clean waters. The region below the limit of light penetration is still the benthic environment but is not usually populated by any photosynthetic plant.

The benthic algae are often light limited and grow slowly due to shading or turbidity produced either by planktonic algae, aquatic vascular plants, terrestrial plants, the wave action, the land runoff or the animal activity. The benthic algae receive light which has been spectrally altered by its passage through water and overlying phytoplanktons. The dense growth of phytoplanktons will considerably reduce the depth of benthic algal occurrence. The mechanical stress of wave action in shallow exposed area can also be a limiting factor to the growth of benthic algae and when they are covered with even less than 1 mm of muddy sediments, their growth is sharply reduced.

In the densely populated benthic community, the competition for space provided with sufficient light is likely to be severe and the biotic interactions among algae, bacteria, fungi, higher plants and grazing animals becomes much more complex than they are in the open water. The heavy settlement of the algal spores may lead to competition between germlings and there is an inverse relationship between the density and individual size. The benthic algae occurring in eulittoral zone are subject to extreme fluctuations in environmental parameters like temperatures and water stress (to some extent) than those occurring in sublittoral region, but, nutrients are not considered to be a limiting factor for the growth of benthic algae at any time since nearby sediments with their abundant heterotrophs contribute enough dissolved nutrients to benthic algae. Some of the macroscopic benthic algae like *Cladophora* forms zoospores, which are planktonic for sometimes in the water. Some benthic algae like *Cladophora, Spirogyra, Mougeotia, Oedogonium* etc. can stir up into overlying water by water currents and can greatly exceed the number of phytoplanktons.

Classification

Depending upon the nature of substratum to which they are attached, they are broadly classified as *epilithon* (those attached to rock), *epipsammon* (those attached to sand), epipelon (those attached to mud), *epiphyton* (those attached to other plants), *metaphyton* (those loosely associated with attached algae or higher plants) and *epizoon* (those attached to animals).

2.2.1 Epilithon

They are those algae which are attached to rock surface in the littoral zone and are present where the water movement is sufficient in order to prevent the accumulation of fine sediments on the rock surface. But if the water movement is slight and the rock surface is flat, it will soon be covered with silt and the epilithic flora may give way to an epipelic flora. It is only the steep rock surfaces or an outstanding rock slopes which are most favoured for colonization by epilithic algae. The angle of slopes and the presence or absence of cracks and grooves on the rock surface can also act as modifying factors for the occurrence of benthic algae. The presence of fissures, cracks, scratches and rough areas on the rock surface permit rich growth of epilithic algae because of the trapping of the reproductive cells. Since the rock is a relatively fix substratum, a permanent epilithic algal flora exist.

The algae are firmly attached to rock surface either by (i) secretion of a slimy gelatinous cushion, sheath, stalk or tube as in diatoms and blue-green algae, (ii) prostrate filaments which support erect filaments such as in green alga *Stigeoclonium*, (iii) holdfast cells as in green algae *Ulothrix* and *Oedogonium* or (iv) by rhizoidal extensions of the basal cell as in *Cladophora*.

A rich algal flora firmly attached to rock surface is usually dominated by blue-green algae (*Chroococcus, Aphanocapsa, Hapalosiphon, Nostoc, Calothrix, Gloeocapsa, Scytonema, Tolypothrix, Rivularia, Stigonema, Schizothrix, Lyngbya, Phormidium* etc.), green algae (*oedogonium, Chaetophora, Ulothrix, Stigeoclonium, Cladophora, Pithophora, Spirogyra, Mougeotia* etc.) and diatoms (*Achnanthes, Diatoma, Synedra, Meridion, Tabellaria, Cymbella, Eunotia, Gomphonema, Peronia, Frustulia* etc.) (Fig. 2.9).

2.2.2 Epipsammon

They are those algae which are attached to sand particles at lake margins and they occurs only on sand grains which are relatively undisturbed by drifting wave action. Algae usually colonizes the hollows or depressions of the sand particles, where they are somewhat protected from abrasion or grinding movement of the sand particles. Sandy substratum rarely has any macroscopic algae attached to it. The epipsammic algae helps in cementing the sand grains by secretion of mucilage or calcium carbonate. The epipsammic flora is mostly consisting of diatoms (*Achnanthes, Amphora, Opephora, Nitzschia, Cymbella, Navicula* etc.) and some small coccoid and filamentous blue-green algae (*Chroococcus, Aphanocapsa, Oscillatoria, Phormidium, Calothrix, Lyngbya* etc.). They are either closely adnate to the sand surface or are raised on very short mucilage stalk. The number of epipsammic diatoms is positively correlated with the amount of organic matter present on the sand grains and negatively with the decrease in light intensities with increasing depth. The diatoms occurring on the sand particles use their raphe system more as an attachment organelle than as motility organelle, but some may move rather slowly from grain to grain to colonize new grains.

2.2.3 Epipelon

The epipelic flora consist mostly of algae living in or on the surface of soft muddy sediments to which light penetrates. They can not withstand long periods of darkness and anaerobic conditions produced either by water movement or activities of numerous animals living in or on the sediment, and only those algae which are capable of positive phototactic movement are likely to survive. The epipelic algae are best developed on relatively stable sediments and can creep over and in between the silt particles in the upper few millimetres of interstitial water. Several epipelic diatoms secrete mucilage which can consolidate and stabilize the sediments.

The epipelic flora is dominated by pennate diatoms (*Pleurosigma, Caloneis, Diploneis, Frustulia, Gyrosigma, Cymatopleura, Neidium, Navicula, Pinnularia, Nitzschia, Stauroneis, Surirella* etc.) and blue-green algae (*Aphanocapsa, Aphanothece, Oscillatoria, Nostoc, Spirulina* etc.), but some members of green algae (*Chara, Nitella, Spirogyra, Mougeotia, Zygnema, Ulothrix, Microspora, Oedogonium, Closterium, Cosmarium* etc.) as well as *Vaucheria* are also encountered (Fig. 2.10). Most of the blue-green algae and green algae exist as loose mucilage mass and lie on the surface of muddy sediment, but some algae like *Chara* and *Nitella* are anchored to silty or muddy sediments by means of branched rhizoids.

2.2.4 Epiphyton

An algal flora attached to other macroscopic algae or higher plants (called as basiphytes) are called epiphyton (the periphyton is superfluous). Depending upon the attachment of epiphyton on basiphytes leaves, stems or roots, they are called as epiphyllous, epiphloeophyllous or epirhizophyllous, respectively. An epiphytic production can not be more than 1-2% of the total primary production of lake even when it is a large shallow one. Most epiphytes have little host specificity. There is an evidence of reciprocal transport of nutrients between epiphytes and basiphytes, but the main pathway seems to be from basiphyte to epiphytes. The epiphytism for an alga appears to be a consequence of competition for substrate, light and nutrients in a crowded benthic environment or it is possible that an alga may be an obligate epiphyte attached only to narrow range of basiphytes.

Life forms of epiphyton The epiphytes may have two distinct forms of life.

(a) *Adpressed type*: They are epiphytes which grow in direct contact with the basiphyte and sometimes coat the entire surface of the basiphyte. They are very much subject to secretions of basiphyte. Some of the adpressed epiphytes are diatoms (*Cocconeis, Amphora, Epithemia, Eunotia* etc.), green algae (*Coleochaete, Aphanochaete* etc.) and blue-green algae like *Xenococcus* (Fig. 2.11).

(b) *Pedunculate type*: They are those epiphytes which grow in an extended habit, with only their basal cell or a basal mucilage stalk in contact with the basiphyte. The pedunculate epiphytes are subject to water flow and receive little nutrients, inhibitors etc. from the host. They occur most probably in a relatively non-turbulent water. The pedunculate epiphytes are much more subject to losses from mechanical and grazing influences than adpressed types. Some of the pedunculate epiphytes are diatoms (*Gomphonema, Synedra, Tabellaria, Cymbella, Achnanthes* etc.) and green algae (*Oedogonium, Bulbochaete, Stigeoclonium* etc.) (Fig. 2.11.)

Factors influencing the epiphytic colonization of basiphytes: The epiphytic colonization of basiphyte is dependent upon several factors like the availability of optimum levels of light, temperature and nutrients; presence or absence of water; basiphyte age, its surface properties and its place of occurrence; and the availability of sporulating stage of epiphytes. The basiphytes at or above the water surface have a sparse epiphytic flora, which varies positively with the degree of wetting by spray than those which are immediately below the water level. A decrease in epiphytes has also been observed in deep water due to dim light. A greater number of epiphytes occur on the upper surface of basiphyte than on its under surface, because of the greater availability of light on the upper surface. The different parts of a larger basiphyte plant are subjected to quite different current regimes, light intensities etc. and may afford more than a single niche for colonization by different epiphytic species. For example, on a single plant of *Cladophora*, there is an apical-basal differentiation of epiphytic distribution. The diatoms *Synedra* and *Licmophora* usually colonizes its upper branches, while *Rhoicosphenia* and *Epithemia* near its base and *Cocconeis* equally all over.

The diatoms *Epithemia* and *Rhopalodia* are common as epiphytes especially in base-rich waters, while diatoms *Tabellaria, Eunotia* and *Actinella* and green algae *Oedogonium* and *Bulbochaete* are usually common as epiphytes in slightly acidic waters. In this century, an increase in epiphytic loading has occurred mainly due to changes in agricultural practice and the influx of tourists, sewage etc.

The fact that epiphytes tend to develop most abundantly in axils, constrictions, grooves or close to mid-rib of basiphyte suggest the trapping of spores and protection of germlings at such places. In general, the degree of colonization is positively correlated with the age of basiphyte and the protection from wave action. The new leaves and stems of basiphytes are naturally free from epiphytes, but as they grow they are colonized by epiphytes and the older leaves and stems of basiphytes are almost obscured by the epiphytic growth. The epiphytic productivity and biomass both are at maximum during summer and autumn than in other seasons.

Influences of epiphytes on basiphytes: The excessive development of epiphytes on basiphyte adversely affect the photosynthesis of basiphyte by producing the light attenuation and reducing the bicarbonate availability. The additional weight of epiphytes must cause a considerable strain on the basiphyte and may be one of the factor responsible for shedding of basiphyte plant parts. The epiphytes may also compete with basiphyte for nutrients like phosphorus etc. present in the water.

Adaptations in basiphytes to avoid epiphytic coating : Certain basiphytes develops certain adaptations in order to avoid epiphytic attachment , e.g. (i) the highly mucilaginous walls of many members of Zygnemaphyceae are relatively free from epiphytes, because many epiphytes are unable to attach to a mucilaginous surface, (ii) the extracellular tannins produced by many members like *Spirogyra, Mougeotia* etc. have antibiotic properties and prevent spore attachment and germling growth of epiphytes. Some basiphytes like *Oedogonium, Bulbochaete, Cladophora* etc. have no mucilage and tannin production and are reachly coated with number of epiphytes, (iii) several basiphytes slough off their outer surfaces and reduces epiphytic loading, and (iv) Charophycean members are often coated with calcium carbonate and which may be a factor in reducing epiphytic flora.

2.2.5 Metaphyton

The metaphyton was first proposed by Behre in 1956. The metaphyton are those algae which are not actually attached to a substratum but are found loosely associated with any benthic algae, higher plants, their epiphytes or debris in shallow water. They are mostly non-motile or slightly motile algae living in the mass of mucilage secreted by the host plants to which they are associated and can be washed out easily by washing the host plants with jets of water, but epiphytes are often found to be so rigidly attached to host plants that only scraping with a sharp scalpel will remove them. This flocculent algal mass with little relationship to the hcst plant surface is involved in the metaphyton.

The origin of metaphyton always seems to be from cells which commence growth adjacent to the host. The metaphyton is richest of all communities developed in a lake. The metaphyton is common at low rates of water flow, since movement of water tend to remove the loosely attached metaphyton and wash it out into phytoplankton.

The metaphyton is usually dominated by several green algae like *Scenedesmus, Oocystis, Cosmarium, Euastrum, Closterium, Staurastrum, Ulothrix, Mougeotia, Spirogyra, Zygnema*; some diatoms like *Pinnularia, Frustulia, Cymbella, Navicula, Tabellaria*; and a blue-green alga *Oscillatoria* and a dinoflagellate *Gymnodinium.*

2.2.6 Epizoon

Some algae are attached to hard shells of less motile sedentary animals and are benefited by being attached to live animals, because they will now be protected by grazing animals and will also obtain some growth stimulants in the form of waste products of animals. The filaments of Stigeoclonium are attached to molluscan shells, while that of *Basicladia, Dermatophyton* and *Oscillatoria* are attached to freshwater turtles.

2.2.7 Seasonal pattern of growth in benthic algae

Only rarely has the succession of benthic algae during an annual cycle has been studied. Like planktonic diatoms, benthic diatoms *Diatoma Vulgare, Cymbella ventricosa, Gomphonema olivaceum* etc. also develops bloom in the spring. But the actual reason for the bloom formation of benthic diatoms is not known, since similar bloom formation by planktonic diatoms at that time produce shade to the benthic diatoms and nutrients shortage never develops in any benthic habitats. The filamentous green algae like *Cladophora, Bulbochaete, Spirogyra* etc and some blue-green algae overgrow the diatoms in mid-summer and supress their growth. There may often be an another smaller peak of diatoms growth in autumn. The decreased population of benthic algae in winter coincides with the low levels of light.

Table 2.1 Categories of green algae represented in freshwater plankton

Unicellular		*Colonial*		*Filamentous*
Motile *(by flagella)*	*Nonmotile*	*Motile* *(by flagella)*	*Nonmotile*	*Motile* *(by mucilage secretion)*
Chloromonas (2 μm)	*Chlorella* (1-2 μm, 10-15 μm, spherical)	*Gonium* (4 × 32 cells, ca. 20 μm plate like quadrangular)	*Ankistrodesmus* (cell no. variable, 2-50 μm)	Desmids
Chlamydomonas (2-100 μm, biflagellate)	*Ankistrodesmus* (2 × 10 μm, acicular, crescent shape)	*Eudorina* (16-32 cells, ca. 20-30 μm, colony ovate)	*Kirchneriella* (4-16 cells, 2-20 μm, sickle-shaped cells)	*Desmidium* (many cells, ca. 120 μm per cell, filament spirally twisted)
Carteria (10-100 μm. quadriflage-llate)	*Stichococcus* (2-8 μm, wide, short cylindrical cells)	*Pandorina* (16-32 cells, ca. 25 μm, cells compactly arranged with trucate apices)	*Scenedesmus* (2-4 cells, 4-50 μm, cells side by side)	Zygnemataceae *Mougeotia* (many cells, bar-like rotating chloroplast)
Chlorogonium (5-100 μm, biflagellate, cells elongate)	*Chlorococcum* (10-20 μm, spherical)	*Volvox* (1-50 × 10^3 cells, ca. 10-15 μm, spherical colony)	*Dictyospherium* (8-32 cells, 4-10 μm, cells in clustures at the ends of radiating strands)	*Zygnema* (many cells, two stellate chloroplasts)
	Desmins (30-300 μm)		*Pediastrum* (8-64 cells, 5-40 μm)	*Spirogyra* (many cells, spiral chloroplasts)
	Closterium			
	Cosmarium			
	Euastrum			

Table 2.2. Members of blue-green algae commonly represented in fresh water plankton.

Genus	*Salient characters*
Microcystis	Cells spherical, 1-9 µm in diameter, ellipsoidal, crowded and not evenly distributed in the colourless mucilage; colonies first rounded later irregular; cell division in all direction; gas vacuoles often present.
Synechococcus	Cells ovate or oblong, sometimes 2 to 4 cells seriately united as a result of cell division in one plane; cells without a sheath; cell content pale blue-green, highly granular.
Aphanocapsa	Cells spherical, 1-6 µm in diameter, evenly and remotely distributed within the globular, ovate or amorphous mucilage mass; cells with a thin gelatinized individual sheath; gas vacuoles wanting.
Aphanothece	Cells ellipsoidal to cylindrical, 1-7 µm in diameter and 2 to 3 times the diameter in length, scattered throughout firm and copious mucilage, cell sheath indistinct and confluent with the homogeneous colonial mucilage; cell division transverse.
Merismopedia	Cells 4-16 (or more), 1-7 µm in diameter, in a plate like tabular colonies, arranged in a single plane; cells globose, division in two directions.
Coelosphaerium	Cells spherical or hemispherical, 1-7 µm in diameter, forming a hollow spherical colonies; colonial mucilage colourless; colonies dividing by breaking through in the middle; gas vacuoles may be present in the cells.
Arthrospira	Trichomes multicellular, cylindrical, without sheath, 2.5-8 µm in diameter, loosely and irregularly coiled with comparatively short and fewer coils, cross wall distinct.
Spirulina	A spirally twisted, unicellular cylindrical trichome without a sheath; spiral loose or close and tightly coiled, very regular.
Oscillatoria	Trichomes filamentous, 0.6-80 µm in diameter, without a sheath; mature trichomes showing a polarity, apical cell sometimes with a calyptra; microscopically usually showing an oscillating or gliding movement.
Lyngbya	Trichomes mostly cylindrical throughout with a firm prominent colourless sheath.
Anabaena	Trichomes mostly gregarious, much entangled and enclosed in amorphous mucilage; they are uniformly broad throughout, sheath-absent or diffluent, heterocyst intercalary; spores single or in series.
Aphanizomenon	Trichomes without sheath, straight, in bundles; heterocyst cylindric, spore cylindrical, single.

Table 2.3 Members of diatoms commonly represented in freshwater plankton.

Genus	*Salient features*
Asterionella	Frustules in girdle view elongate rectangular, forming a circular colony in which the cells radiate from a common centre like spokes of a wheel, in valve view slightly enlarged at the poles.
Fragilaria	Frustules quadrate or rectangular in girdle view, form ribbons; valve view fusiform, the poles narrowed from an enlarged central region.
Stephanodiscus	Valves unevenly ornamented, the radiating rows of puncta separated by clear, smooth, radiating zones; intramarginal circle of coarse spines which extend beyond the edge of the valve.
Tabellaria	Frustules attached in zig-zag chains; longitudinal septa present, straight; rows of transverse puncta visible in valve view.
Cyclotella	Valves with an intramarginal zone of costae, smooth or finely punctate within the coastal zone; with markings in the centre zone or smooth.
Melosira	Frustules cylindrical in girdle view, attached to form filaments; polar margins often with denticulations.
Diatoma	Valve view symmetrical, usually elliptic or subcylindric, often with subcapitate poles; in valve view with a faint pseudoraphe, girdle view rectangular.

Table 2.4 : Chrysophytes members commonly represented in freshwater plankton

Naked cells	*Cells covered with silicified scales and bristles*	*Cells covered by an organic lorica*
Chromulina	*Mallomonas*	*Chrysococcus*
(Solitary cells, a single apical pleuronematic flagellum, one or two chloroplast)	(free-swimming cells; two flagella, one is reduced, needle like projection borne on scales)	(structure like *Chromulina,* protoplast is enclosed within a spherical or ovoid lorica possessing a narrow apical aperture for the protrusion of the long flagellum)
Ochromonas	*Synura*	*Dinobryon*
(Viflagellate cells, one or two chloroplasts, shows auxotrophy for various vitamins, posterior end of cell gradually tapering to a point)	(spherical colonies, cells held together by their elongated posterior ends, cells bears two heterodynamic flagella of unequal length)	(biflagellate cells resembling *Ochromonas,* forming arborescent colonies, loricas are broad and open vase-shaped structures, cells attached to inside lorica by a narrow cytoplasmic strand, two to four cellular progeny are produced by cell devision.)
Uroglena	*Chrysosphaerella*	
(*Ochromonas* type cells, cells are grouped at the periphery of a spherical or ellipsoidal mass of mucus, the individual cells are connected by a system of thread - like, forking mucilage - strands emanating from the centre of the colony)	(globose colony of ellipsoid or pyriform cells, the cells furnished with 2 long anterior rods set in basal cups, one flagellum arising from between the two sheathed setae, chloroplast 2)	

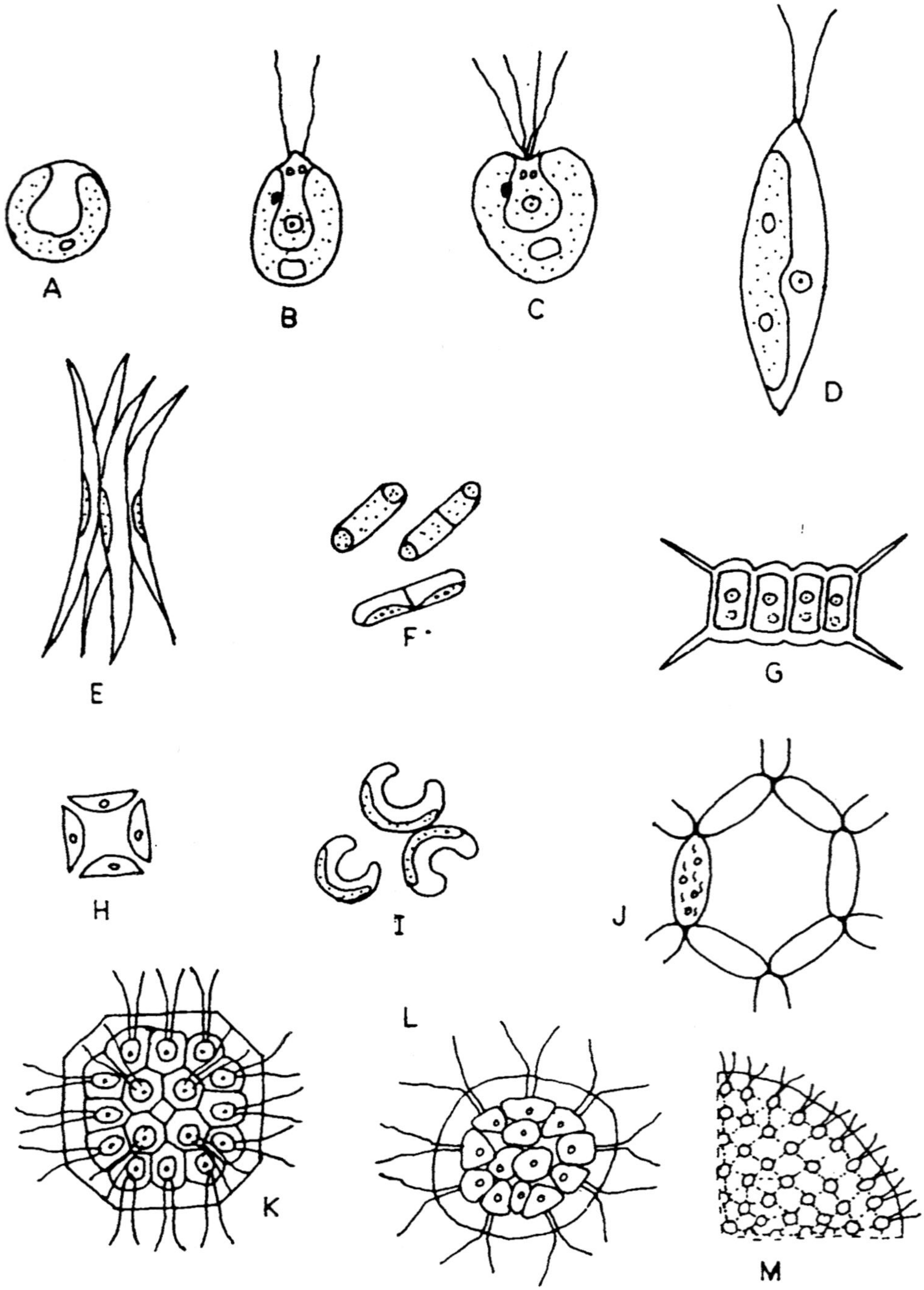

Fig. 2.1 Some green phytoplankton members. A. *Chlorella*; B. *Chlamydomonas*; C. *Carteria*; D. *Chlorogonium*; E. *Ankistrodesmus*; F. *Stichococcus*; G. *Scenedesmus*; H. *Crucigenia*; I. *Kirchneriella*; J. *Hydrodictyon*; K. *Gonium*; L. *Pandorina*; M. *Volvox*.

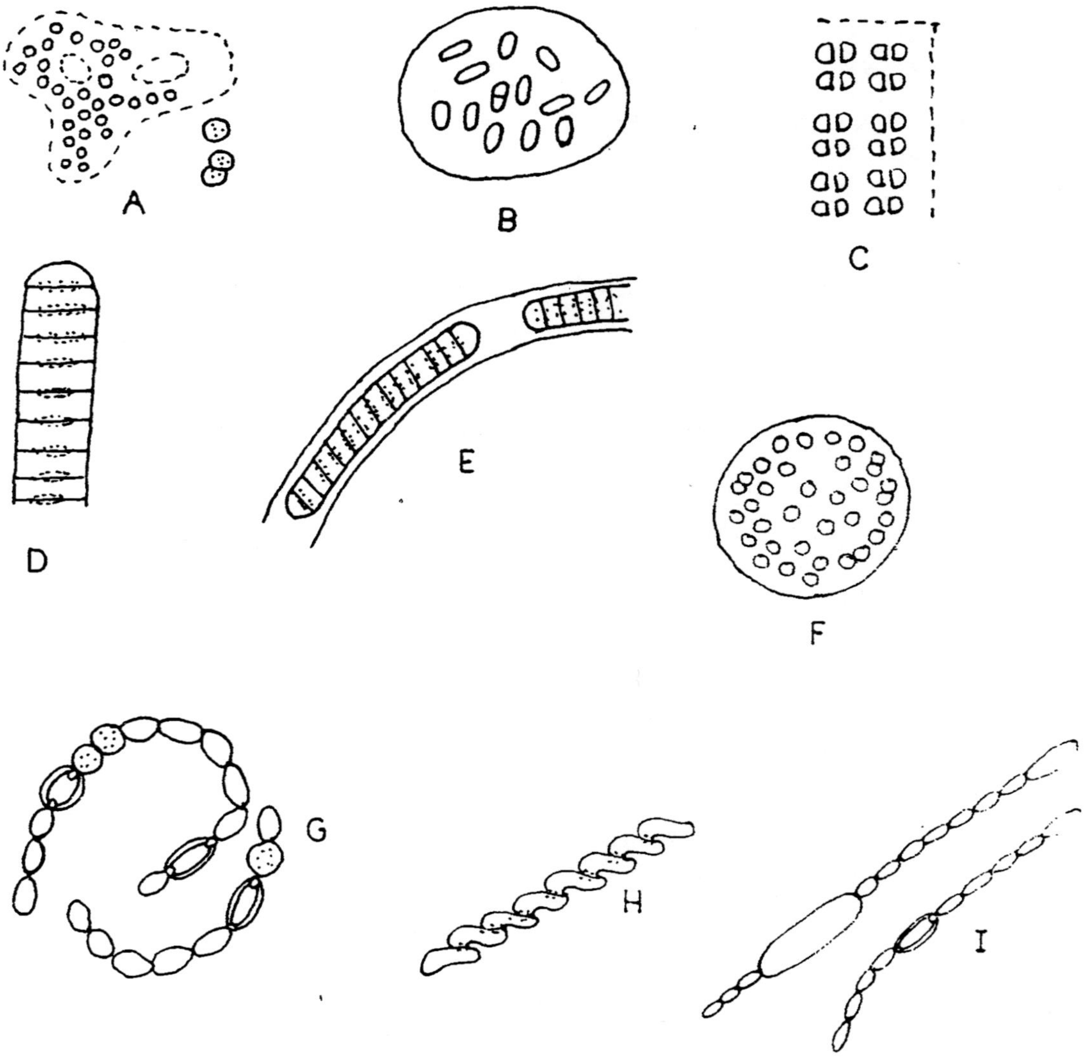

Fig. 2.2 Showing some members of planktonic blue-green algae. A. *Microcystis*; B. *Aphanothece*; C. *Merismopedia*; D. *Oscillatoria*; E. *Lyngbya*; F. *Coelosphaerium*; G. *Anabaena*; H. *Spirulina*; I. *Aphanizomenon*.

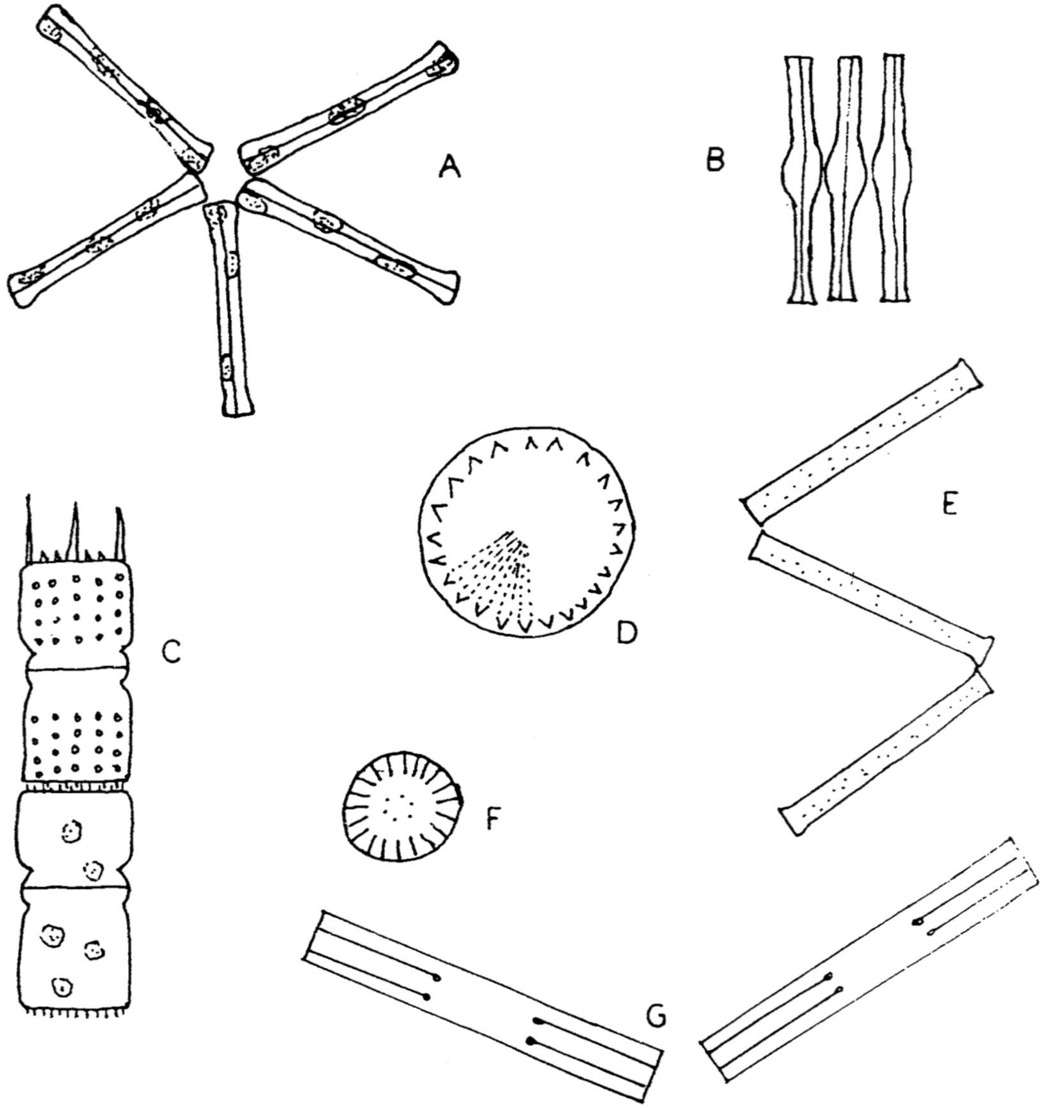

Fig. 2.3 Prominent planktonic diatoms found in freshwater. A. *Asterionella*; B. *Fragilaria*; C. *Melosira*; D. *Stephanodiscus*; E. *Diatoma*; F. *Cyclotella*; G. *Tabellaria*.

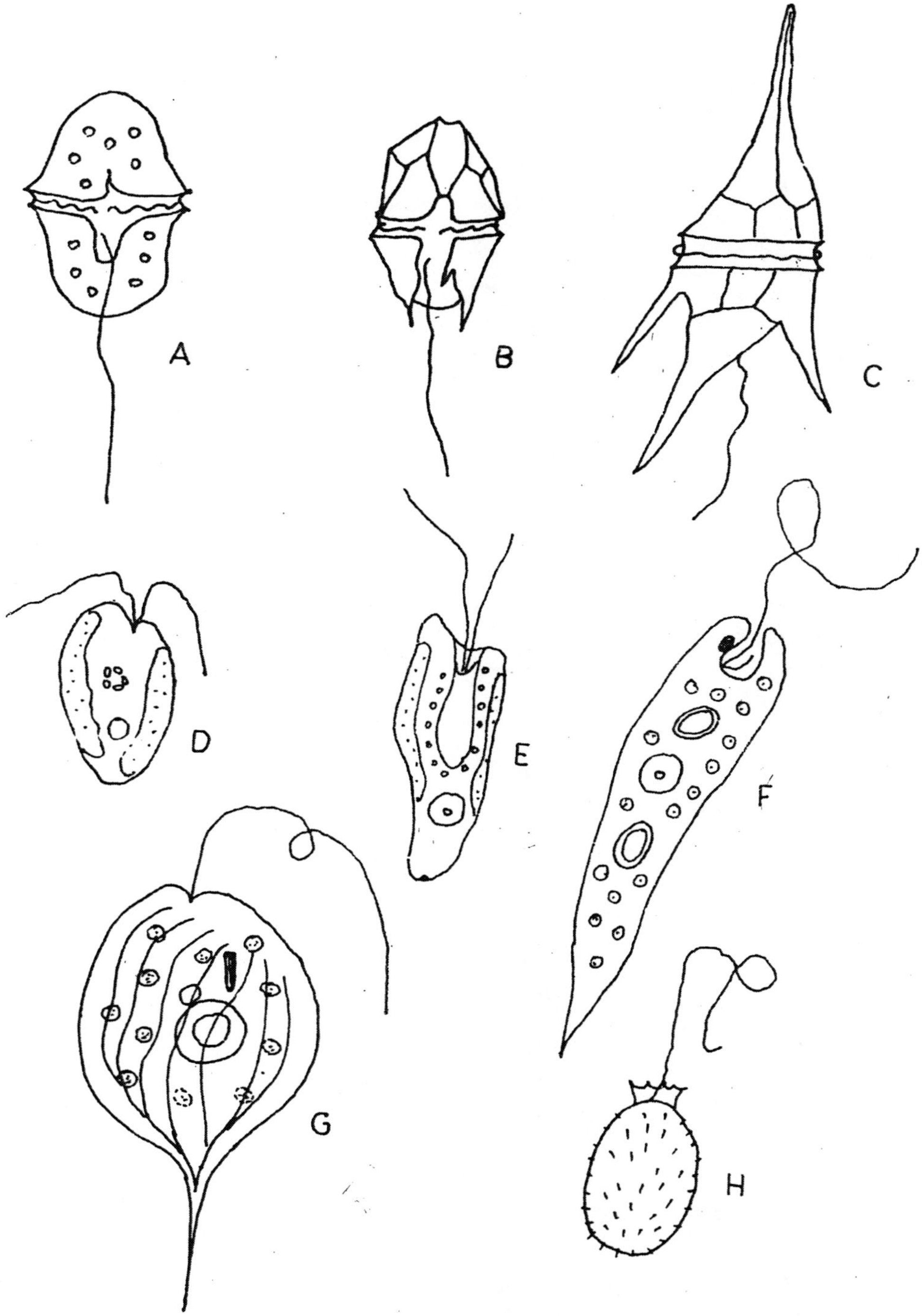

Fig. 2.4 Prominent planktonic dinoflagellates, cryptophytes and euglenoids.
A. *Gymnodinium*; B. *Peridinium*; C. *Ceratium*; D. *Chroomonas*; E. *Cryptomonas*; F. *Euglena*; G. *Phacus*; H. *Trachelomonas*.

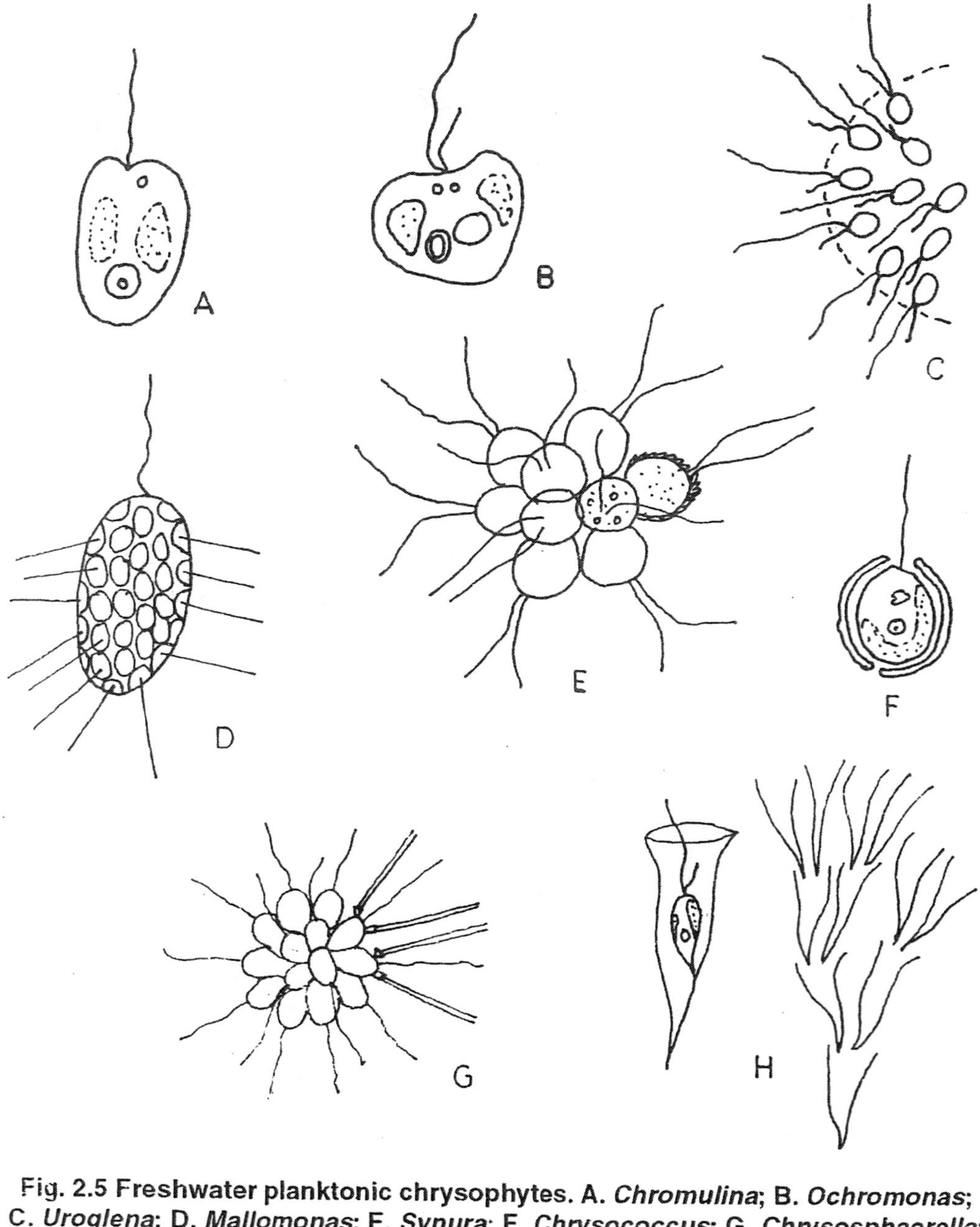

Fig. 2.5 Freshwater planktonic chrysophytes. A. *Chromulina*; B. *Ochromonas*; C. *Uroglena*; D. *Mallomonas*; E. *Synura*; F. *Chrysococcus*; G. *Chrysosphaerella*; H. *Dinobryon*.

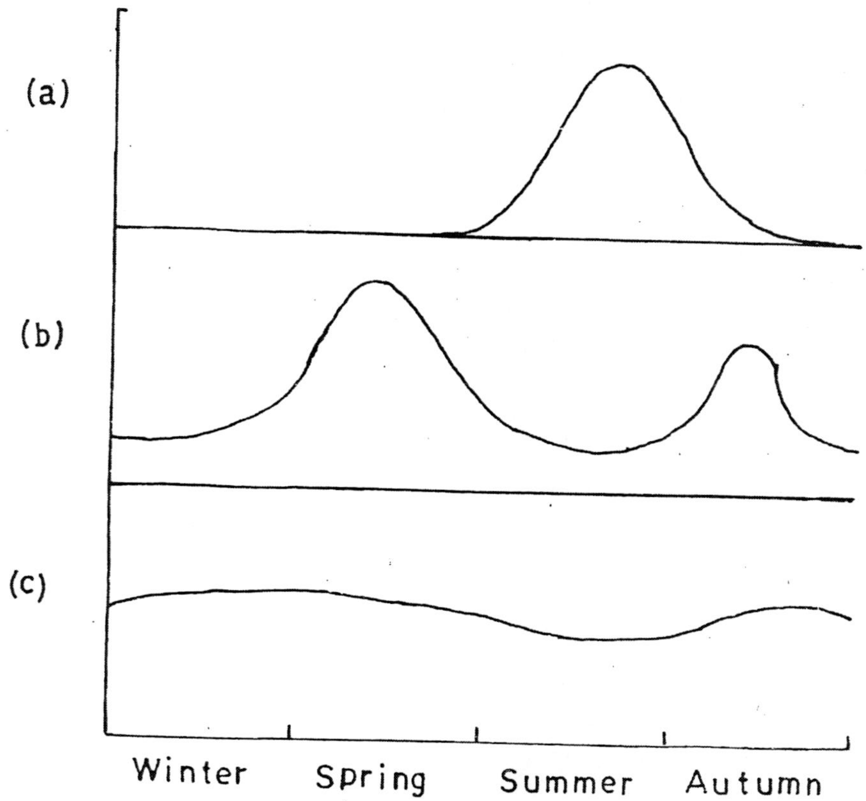

Fig. 2.6 The daily photosynthesis on a cloudless day for the surface water showing the mid-day depression.

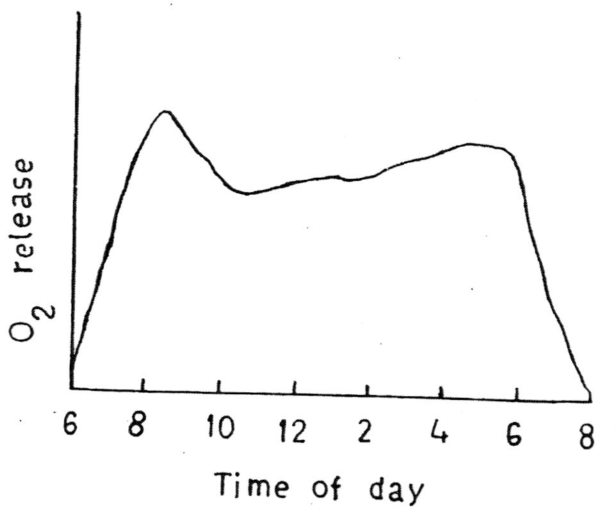

Fig. 2.7 The seasonal amplitudes of phytoplankton production. (a) Polar lakes, (b) Temperate lakes, (c) Tropical lakes.

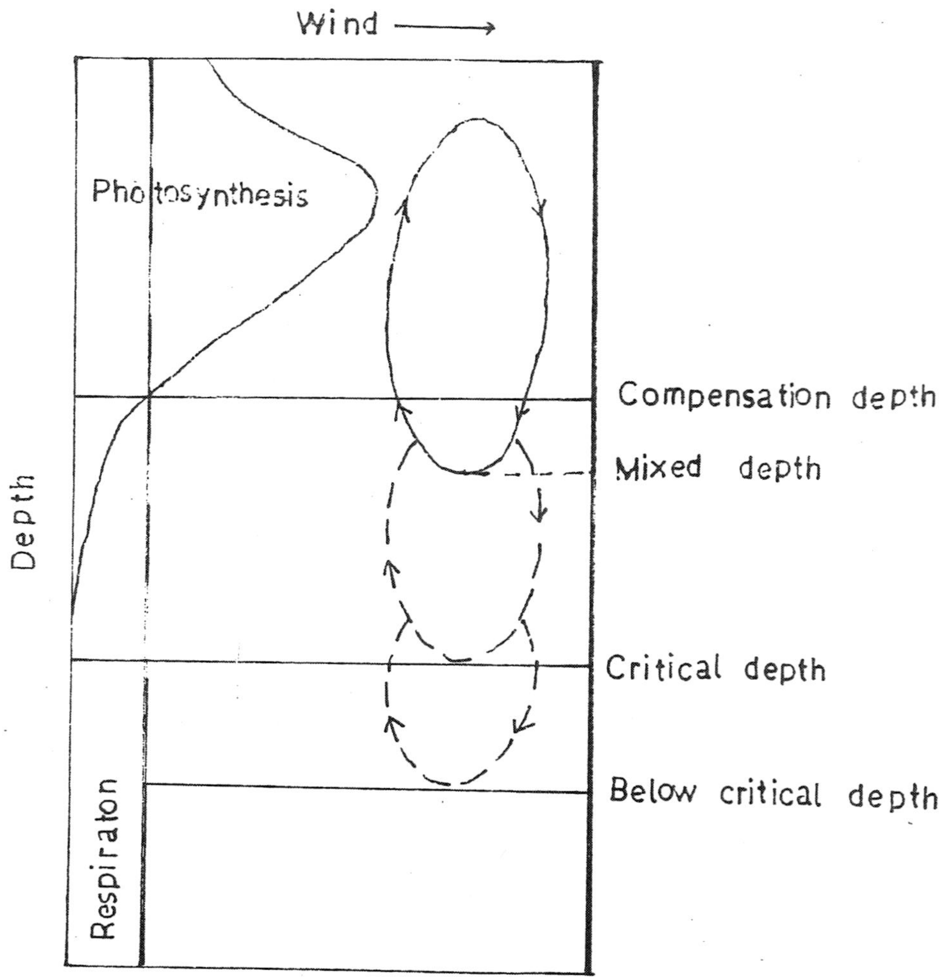

Fig. 2.8 The relationship between the compensation depth and the critical depth. Critical depth is the depth to which the total phytoplankton biomass may be circulated and still spend enough time above the compensation depth to have a total production equal to its total respiration during the same time period.

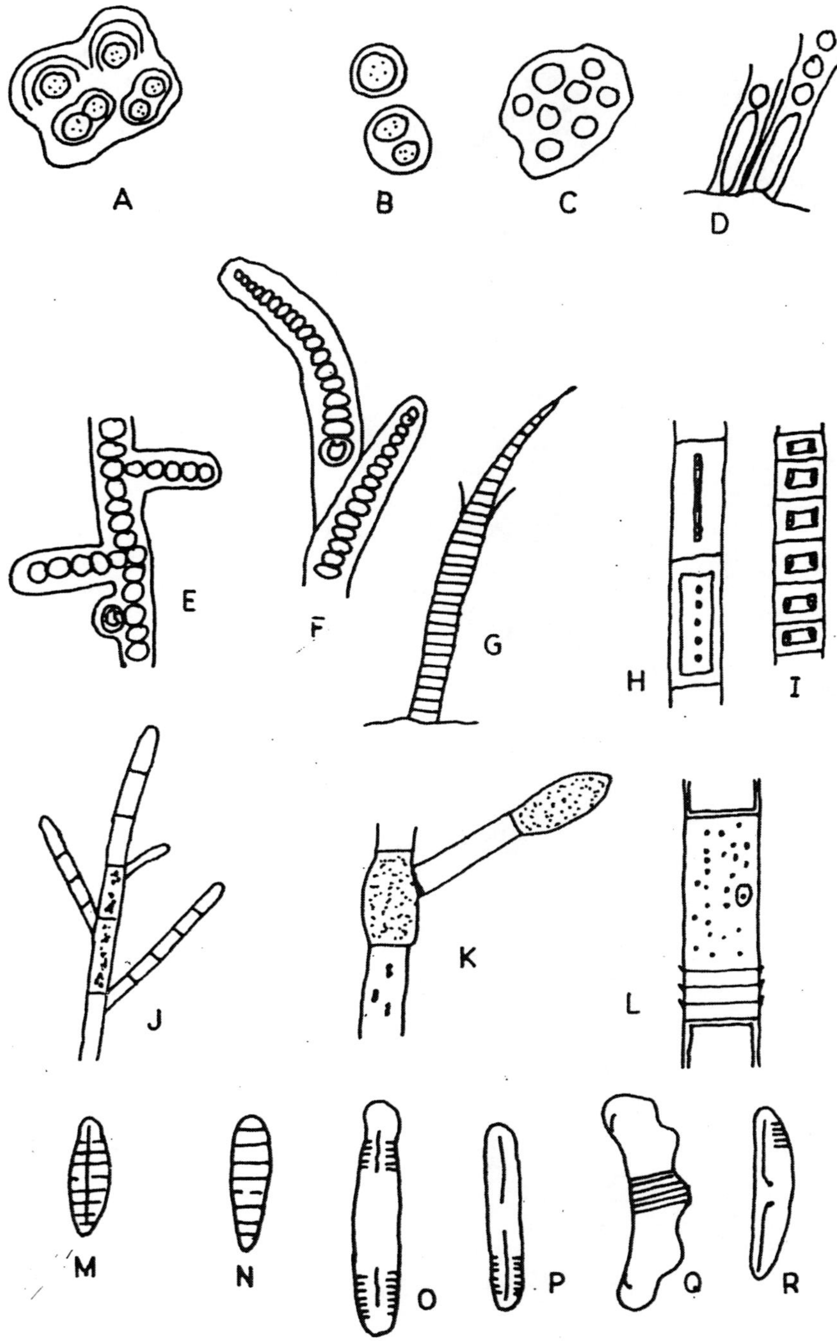

Fig. 2.9 Some freshwater epilithic members. A. *Gloeocapsa*; B. *Chroococcus*; C. *Aphanocapsa*; D. *Chamaesiphon*; E. *Stigonema*; F. *Tolypothrix*; G. *Homeothrix*; H. *Mougeotia*; I. *Ulothrix*; J. *Cladophora*; K. *Pithophora*; L. *Oedogonium*; M. *Diatoma*; N. *Meridion*; O. *Peronia*; P. *Achnanthes*; Q. *Eunotia*; R. *Cymbella*.

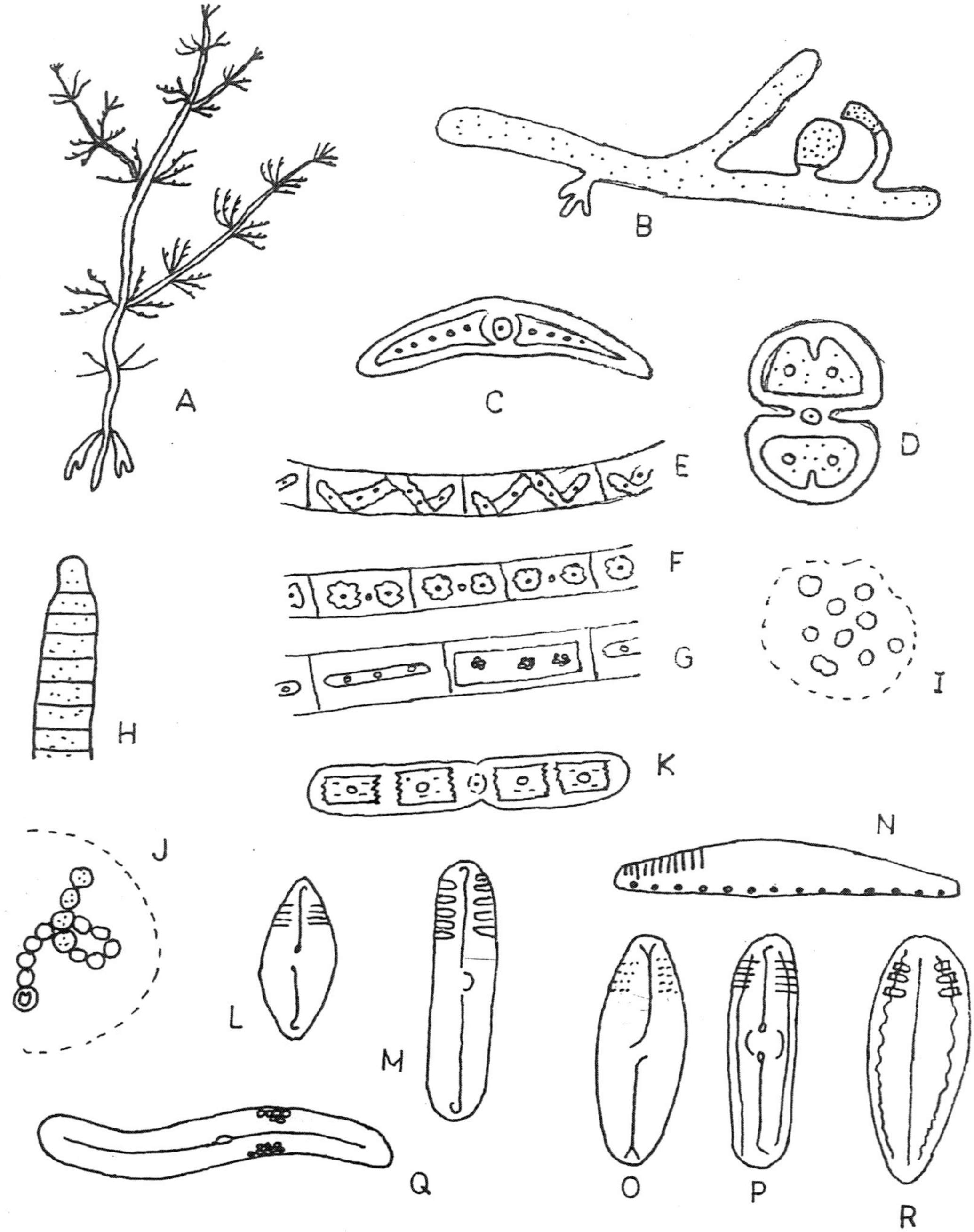

Fig. 2.10 Some freshwater epipelic members. A. *Chara*; B. *Vaucheria*; C. *Closterium*; D. *Cosmarium*; E. *Spirogyra*; F. *Zygnema*; G. *Mougeotia*; H. *Oscillatoria*; I. *Aphanocapsa*; J. *Nostoc*; K. *Netrium*; L. *Navicula*; M. *Pinnularia*; N. *Nitzschia*; O. *Neidium*; P. *Caloneis*; Q. *Pleurosigma*; R. *Surirella*.

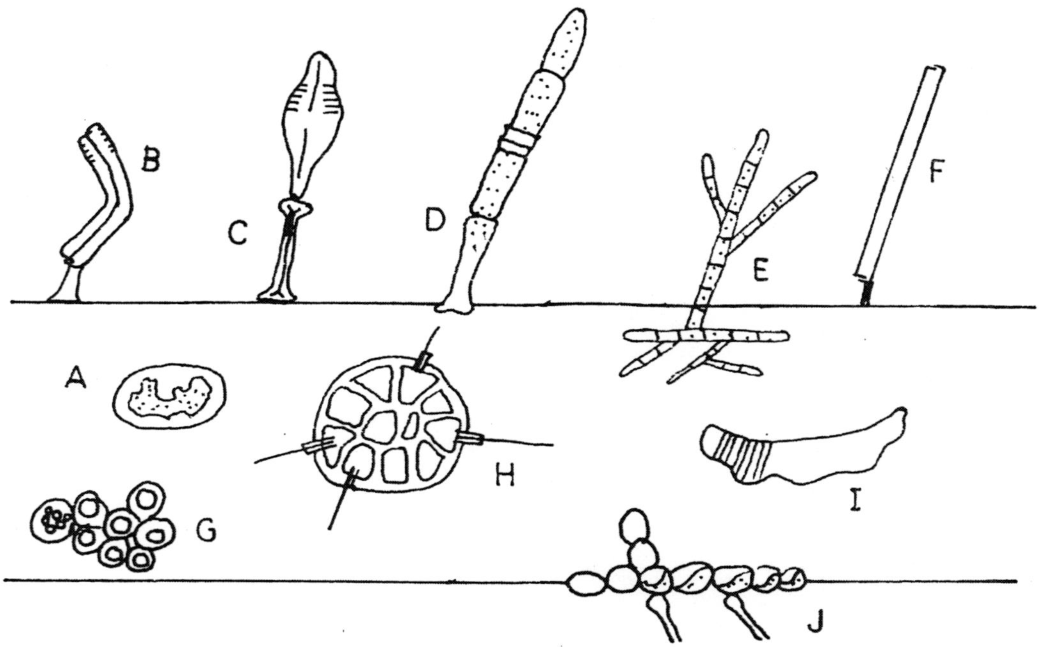

Fig. 2.11 Some freshwater epiphytic members. A. *Cocconeis*; B. *Achnanthes*; C. *Gomphonema*; D. *Oedogonium*; E. *Stigeoclonium*; F. *Synedra*; G *Xenococcus*; H. *Coleochaete*; I. *Eunotia*; J. *Aphanochaete*.

III

LAKE FLORA : HIGHER PLANTS

A small number of aquatic angiosperms and pteridophytes may achieve prominence in some shallow coastal waters of lakes.

3.1 CATEGORIES OF AQUATIC HIGHER PLANTS

The higher plants present in lake can be classified into two main types depending upon their physical relationship with the water and soil.

(i) Free-floating plants

They are not attached to bottom sediment. They drift in the water along with the water current and can be collected at lake margin. They prefer quiet water for their optimum development and can be classified into two main types.

(a) *Surface-floating plants* : They are principally the species of *Wolffia, Lemna, Spirodela, Pistia, Eichhornia and some pteridophytes like Azolla*, *Salvinia* etc. (Table 3.1, Fig. 3.1). They float on the water surface, with most of their body above the water surface. If the roots are present, they hang in the water and are not anchored to the bottom. They move along with the prevailing wind and water current and are cornered in the littoral zone and form a carpet on the water surface. *Eichhornia* is one of the world's worst weed plant known.

(b) *Submerged floating plants* : Some of the prominent submerged floating plants are *Utricularia, Ceratophyllum, Aldrovandia* etc. (Table 3.1, Fig. 3.1). With the possible exception of flowering, they spend their entire life cycle submerged in the water. They have no roots. Their flowers are elevated above the water for insect or wind pollination, except in *Ceratophyllum* where they are submerged and have hydrophilous pollination. The vegetative reproduction is more common in these plants. They perish when exposed to strong illumination. They share the characteristics of delicacy of lamina, absence of a well-differentiated palisade tissue and the presence of chlorophyll in the epidermis. The transpiration stream has no existence in submerged plants.

(ii) Attached plants

They grow in shallow water upto the depth of effective light penetration along the lake margin. Most of them favour mud or mud-sand substratum for their growth and attachment, and are not collected from rocky substratum. Their roots functions more as an organ of anchorage to mud and erectness of flowering stem than for the absorption of water and nutrients. They can be categorized into three different types.

(a) *Rooted plants with emergent leaves* : The plants like *Typha, Sagittaria, Scirpus, Eleocharis, Sparganium, Pontederia* etc, are kept in this category (Table 3.1, Fig. 3.2). They are rooted plants with principal photosynthetic surfaces projecting in the air above the water. They obtain carbon dioxide from the air but nutrients from beneath the water surface. An anchorage of plants to bottom mud or sand prevent them from being swept away by water current.

(b) *Rooted plants with floating leaves* : They are the species of *Nymphaea, Nelumbo, Brasenia* etc., (Table 3.1, Fig. 3.3). Their horizontal photosynthetic leaf lamina floats on the water surface and may more effectively reduce the light penetration into the water than other surface floating plants and algae. Their leaf blades are usually circular and have a tough leathery texture that afford protection against tearing and perforation. During heavy rains, they are liable to be much more severely battered than air leaves which can yield freely in the atmosphere. A leaf with an entire outline is less easily wetted and submerged than the one which is subdivided. It is obvious that the centre of gravity of a floating leaf, which approximates to the circular form, lies at its central point and that this is therefore the most mechanically economical position for petiolar support. The petioles are long, thin and very flexible and their additional length allow the leaf blades to spread out on the water surface, so that they do not overlap each other. Another result of the length and flexibility of the petiole is that the lamina can respond freely to the movement of the water and hence is less liable to be submerged, but in spite of this, the floating leaves still remain liable to serious risk of wetting and submergence and this is proved by the fact that plants bearing such leaves are quite unable to colonize windy and exposed surfaces where the water is liable to be rough.

The upper surface of the leaf is usually covered with a waxy cuticle but the lower surface is devoid of it. The normal stomates are borne upon the upper surface of the floating leaves where they are in contact with the air.

(c) *Rooted plants with submerged leaves* : Some of the important rooted plants are *Potamogeton, Vallisneria, Hydrilla, Myriophyllum, Elodea, Najas* etc., (Table 3.1, Fig. 3.3). The majority of them are characterized by thin elongated, branched stem clothed with leaves and often capable of rooting at the nodes. The submerged stem, like submerged leaves have little or no cuticle on the surface and facilitate the diffusion of dissolved gases and nutrients into and out of the tissue, and contain chloroplasts even in their epidermal cells to absorb dim light in the water.

The submerged leaves in general are thin, usually long either entire or deeply dissected and thus have high surface area to volume ratio for efficient absorbtion of dissolved gases and nutrients from the water. The submerged leaves are poor in stomata and hair, but may commonly bear water stomata, which probably keep up the transpiration stream by exudation. The mesophyll of submerged leaves shows little sign of differentiation into palisade and spongy parenchyma. The assimilatory activity in great measure is confined to the epidermis and the mesophyll serve rather for storage purpose. The epidermis is rich in chloroplasts, while the mesophyll contains large starch grains. The absence of marked dorsiventrality in the leaves of many submerged plants may in part be attributed to the fact that they are perpetually being moved about by water currents, and thus they do not retain any constant position in relation to incident light. The extremely delicate leaf is strengthened by some fibrous cells. The supporting sclerenchyma is characteristic of a certain number of submerged leaves such as those of *Potamogeton.*

In submerged aquatics, the mucilage or slimy coating, which is a by-product of the plants metabolism, may serve to prevent the soluble products of assimilation diffusing into the water or may form a protection against water animals and discourage microparasitic and epiphytic growth.

3.2 ZONATION OF ROOTED HIGHER PLANTS

Under actual field conditions, almost all water plants are limited to the depth of 10 metres or less and are distributed with great regularity. Passing inwards from the shore, the following order is generally observed: (a) there is, firstly, a littoral zone of plants such as *Typha, Sagittaria, Scirpus, Phragmites* etc., standing out of the water, (b) a belt of plants with floating leaves, among which *Nymphaea* is a dominant genus, and still farther from the shore, a zone of plants with leafy shoots reaching to the water surface or nearby, consisting mainly of *Potamogetons* is also observed and (c) a region in which the upper layers of the water are free from the vegetation and the plants lie on the bottom, such as *Ceratophyllum, Elodea, Hydrilla, Chara, Nitella* etc., (the last two are algae). The Figure 3.4 shows essentially similar zones of vegetation in the form of a section.

One of the chief reasons determining this zonation seems to be that plants with floating leaves can only flourish if guarded from the wind and for this reason they generally do not occur at a great distance from the shore and often obtain the necessary protection by growing amongst emergent plants. The submerged plants, as a rule, form a special zone farther from the shore than the floating leaf association, because the latter shades the lower layers of the water so much that the subdued sunlight, that penetrate it, is insufficient to supply a deeper flora.

3.3 SOME PECULIARITIES OF AQUATIC HIGHER PLANTS

(i) World wide distribution

Most of the water plants are widely distributed. The generalization, certainly holds good that aquatic angiosperms have, as a rule, a wider distribution than the terrestrial members of the group. Some species of *Potamogeton, Lemna* and *Ceratophyllum* are found almost all over the world. The dispersal and dissemination of water plants conveyantly takes place through detached fragments and various types of winter buds or turions than fruits or seeds. Wading birds, which are great wanderers, might convey seeds adhering to their feets from one water body to another. The floating species are transported in fragments that adheres to the feet of wading birds. The water plant parts, seeds and fruits are attached to birds feet, beaks or feathers or are present in the form of undigested seeds in the alimentary canal. In damp weather, the plant parts might, for a day or two, withstand exposure to the atmosphere and thus can be carried a few hundred miles entangled in birds plumage.

The water plants shows little change, though the number of species may diminish rapidly as the high altitudes are approached, for example, the plants like *Myriophyllum, Lemna, Isoetes* and *Potamogeton*, which flourishes at sea level have also been found to be present at the heights of 2500-5000 metres.

(ii) Perennial habit

Most of the water plants are perennial. There is of course no dry season to be span and many water plants can continue their vegetative growth all the year round. The perennial habit is directly related to the environment. The conditions under which these plants live, such as unlimited water supply, presence of abundant carbon dioxide and protection from sudden temperature changes are favourable to their growth. The water plants are called upon to endure the smaller range of diurnal and seasonal temperature variations.

(iii) Excessive vegetative reproduction

The excessive vegetative activity of water plants acts in all probability as a deterrent to sexual reproduction. The most distinctive mode of wintering and of vegetative reproduction found among water plants is by means of winter buds or turions. These specialized shoots, which are compact and stored with food materials become detached from the parent plants and pass the winter either floating or resting at the bottom of the water. In the spring, they developed into full-fledged individuals. The water plants may also escape winter by their habit of sinking to the bottom in the cold season.

(iv) Role of intercellular spaces

The existence of a highly developed system of intercellular spaces is one of the most marked anatomical character of the water plants. The air spaces may be formed either by the separation of the adjacent cells or by their destruction. The total volume of oxygen held in the water at ordinary temperatures is extremely small, that in a litre of water, the maximum amount of oxygen present is 10 c.c. as compared with more than 200 c.c. in a litre of atmospheric air. The result is that the water plants have considerably less oxygen at their disposal than is the case with land plants. It is generally assumed that these intercellular spaces or lacular systems serve as store house of oxygen evolved in photosynthesis and its conveyance to the parts of the body that stand in special need of it, more particularly the roots and rhizomes buried in the asphyxiating mud. The mesophyll of the lamina, the ground tissue of the petiole, the cortex of the stem and the roots, all are the regions in which the air spaces reach their greatest development. Due to presence of air in the intercellular space, each stem axis is to a large extent relieved of the task of supporting the weight of its branches. The xylem tends to be reduced in amount, while the lignification is often very poor.

Table 3.1 : Some salient features of dominant water plants.

Plant	*Salient features*
Surface floating plants	
Wolffia	Smallest, rootless flowering plant without any distinction of stem and leaf.
Lemna	Minute flat globular frond with single hanging root which ensure the equilibrium of the plant.
Spirodela	Fronds larger than *Lemna* with numerous roots.
Salvinia	Fern with horizontal floating unbranched stems and without any root; leaves are in whorls of three at each node, two leaves of each whorl are round, floating and photosynthetic with white erect unwettable hairs on their upper surface, while the third one is submerged, non-photosynthetic and divided into numerous hair like segments.
Pistia	Plants consist of floating rosette of pale, yellowish-green, fan-shaped leaves which are deeply ribbed and may be upto 15 cm long; the flowers are small, green and inconspicuous and are born in the centre of the plant among the leaf bases; roots form a fibrous mass beneath the rosette of leaves; vegetative reproduction by stolons from which fresh rosette may arise.
Eichhornia	Plants consist of floating rosette of leaves with inflated, bladder like petioles; roots dark in colour, fibrous and forming a dense mass beneath the rosette; flowers slowy, blue to purple in colour and borne on a spike; plants connected by stolons.
Azolla	Tiny fern with horizontally floating branched stems that are completely hidden by minute overlapping, scale-like leaves that are less than 0.5 mm long, leaves have two lobes, an upper thick green lobe which does not touch the water and a lower, thin, colourless lobe whose bottom surface is in contact with the water, cavities in the upper lobe are inhabited by a blue-green alga *Anabaena azollae* with the ability to fix nitrogen; older plants may become pink, red or maroon in colour.
Free-floating submerged plants	
Utricularia	Thin branched stems totally devoid of roots and covered with delicate hair-like simple or compound leaves divided into hair-like segments, float just

Plant	Salient features
	beneath the water surface; leaves bear tiny bladders which act as traps for small aquatic invertebrates which serve as food for the plant, bladders are interpreted as a modification of part of the leaf; flowers are yellow or purple with one or several born on a thin, wiry scape that protrudes into the air, entomophilous pollination.
Ceratophyllum	Stems long, thin, flexible; plants entirely without roots; leaves flat, forked, divided into several flat, narrow, stiff segments, whorled, more crowded near the tip of the stem; flowers and fruits small and inconspicuous.
Rooted emergent plants	
Sagittaria	They have basal leaves whose blades are usually arrowhead - shape, several slightly curved veins of equal prominence in leaf blades; white flowers borne in 1 to 12 whorls, usually with three flowers per whorl; rootstocks bear tubers which are favourite food of ducks.
Typha	Perennial with stout rhizomes and erect stems upto 10 feet tall; leaves flat, ribbon-like and arise from near ground levels; pistillate portions of inflorescence persist for months and are cigar-shaped.
Scirpus	Rush like plants with stems triangular or rounded; leaves have sheath at their bases, specialized leaves arise from the base of the inflorescence and appear to be in continuation of the stem.
Eleocharis	A tuft or clump of leafless, erect stems arising from subterranean rootstocks; stem round; flowers borne in small, solitary terminal spikelets; stems may be completely submerged in deeper water and than remain sterile.
Phragmites	Tall, perennial grass with creeping stout rhizomes and hollow, round stems which may reach heights of 13 feets; large feathery, plums like panicle.
Rooted plants with floating leaves	
Nymphaea	Leaf blades circular or nearly circular, deeply cleft at the base, float upon the surface of still water, blades attached to subterranean stems by slender flexible petioles; conspicuous flowers floats on the surface, attached to the underground stems by slender peduncles; fruits ripen beneath the water.

Plant	Salient features
Brasenia	It has oval blades which float on the water surface and have no cleft at the base; petioles are connected to the slender, creeping rhizomes by a system of branching stems; flowers small relatively inconspicuous, borne above the water surface.
Rooted plants with submerged leaves	
Elodea	Perennial, slender, green, flexible stems, sparsely branched, rooted to the bottom with fibrous roots; leaves without petioles, arise in whorls of three to seven along the entire length of the stems, leaves strap - shaped; minute white flowers rise to the surface on thread like stalk.
Hydrilla	Identical to *Elodea* but leaf margins have slightly more prominent teeth; plants forms underground tubers and feels harsh and coarse when drawn through hands.
Potamogeton	Very large variable and complex genus, leaves alternate (rarely subopposite) and of two types; floating leaves tough, leathery, while submerged leaves thin, delicate and transluscent, leaf shapes varies from broadly ovate to narrow and hair like; petioles absent or present; stems slender and branched; flowers small, crowded into elongate or globose spike that may be above or below the water surface.
Vallisneria	Long flat ribbon-like leaves upto 1 metre in length arising from stolons or rhizomes, leaf tip bluntly rounded and veins clearly visible; floating female flowers are elongate, tubular and attached to the base of the plant by an extremely long peduncle that coils after pollination, retracting the developing fruit beneath the surface.

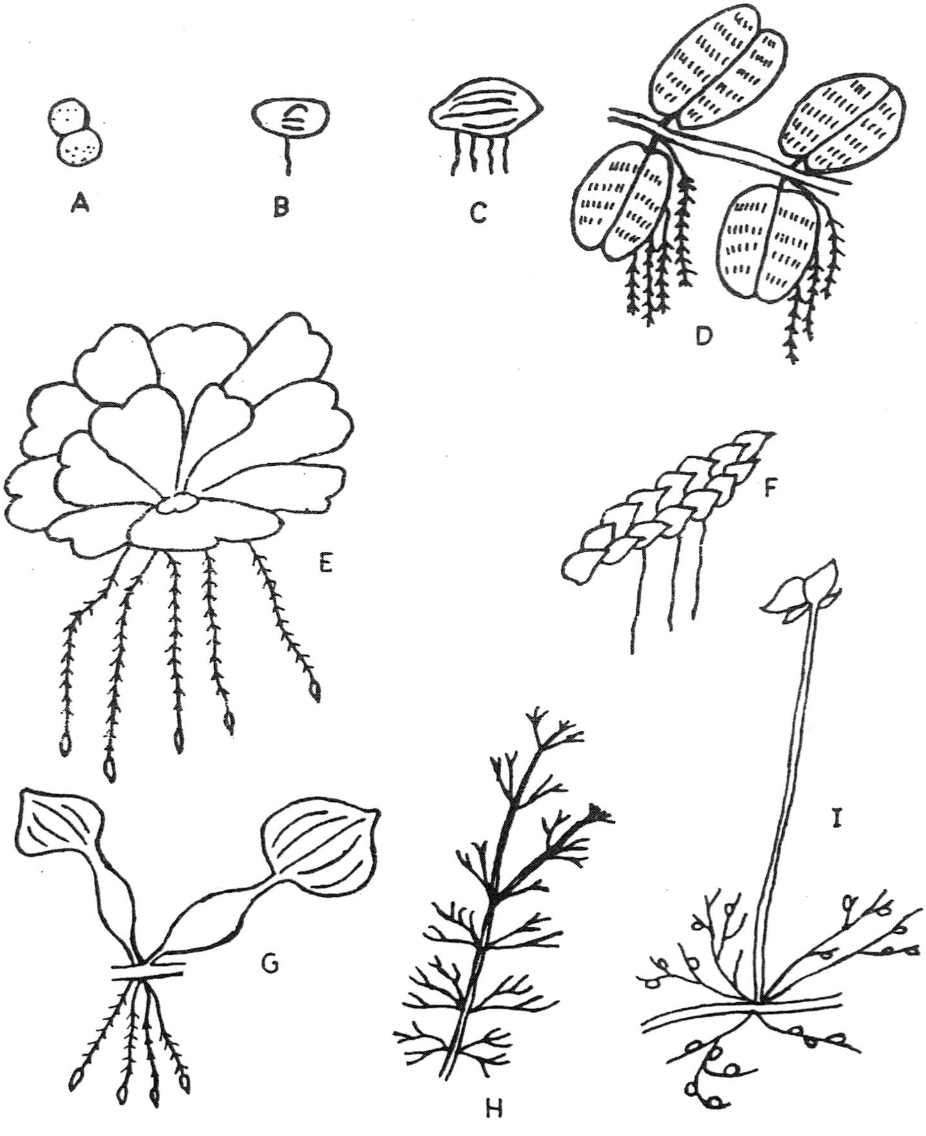

Fig. 3.1 : Some freshwater free - floating higher plants.
A. *Wolffia*; B. *Lemna*; C. *Spirodela*; D. *Salvinia*; E. *Pistia*; F. *Azolla*; G. *Eichhornia*; H. *Ceratophyllum*; I. *Utricularia*.

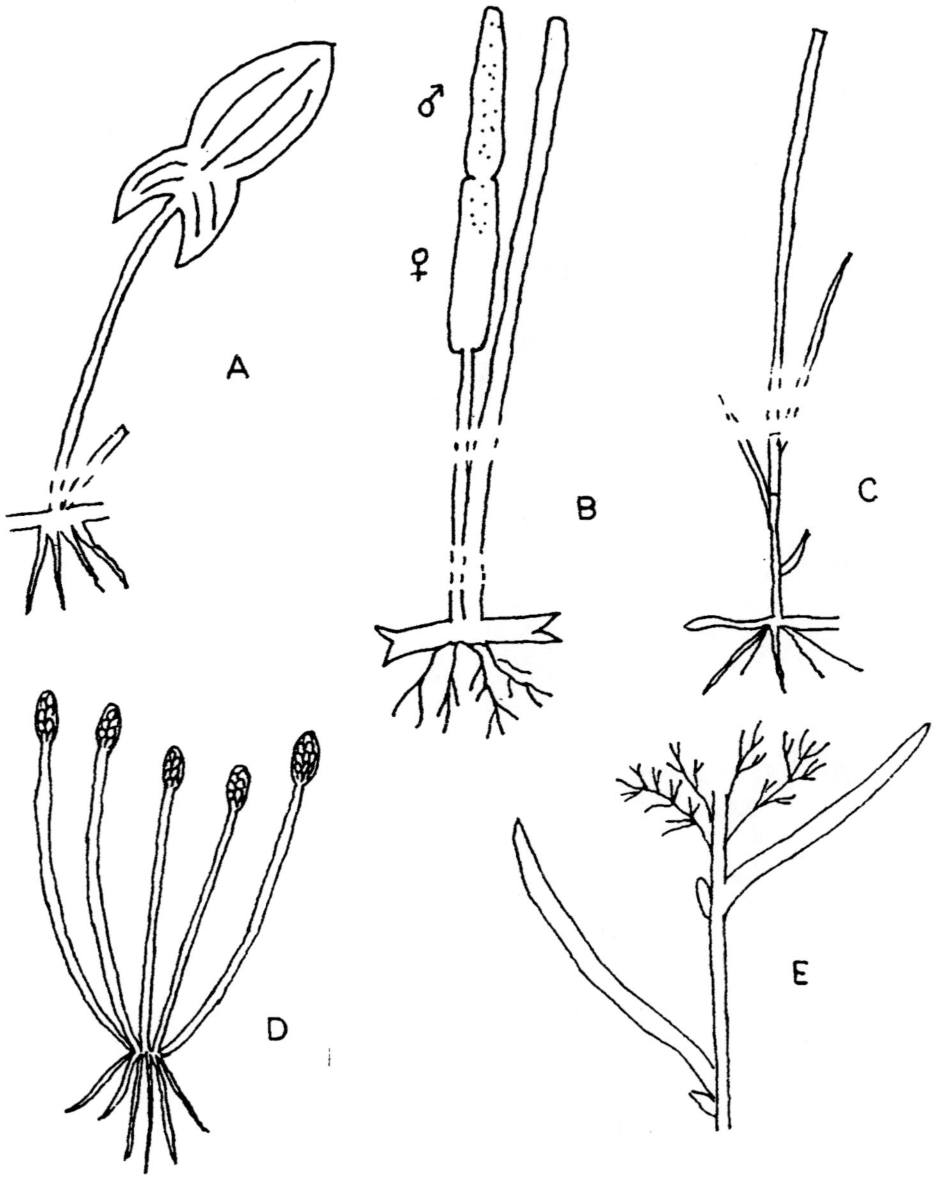

Fig. 3.2 : Some freshwater rooted higher plants with emergent leaves. A. *Sagittaria;* B. *Typha;* C. *Scirpus;* D. *Eleocharis;* E. *Phragmites* (upper portion with part of inflorescence).

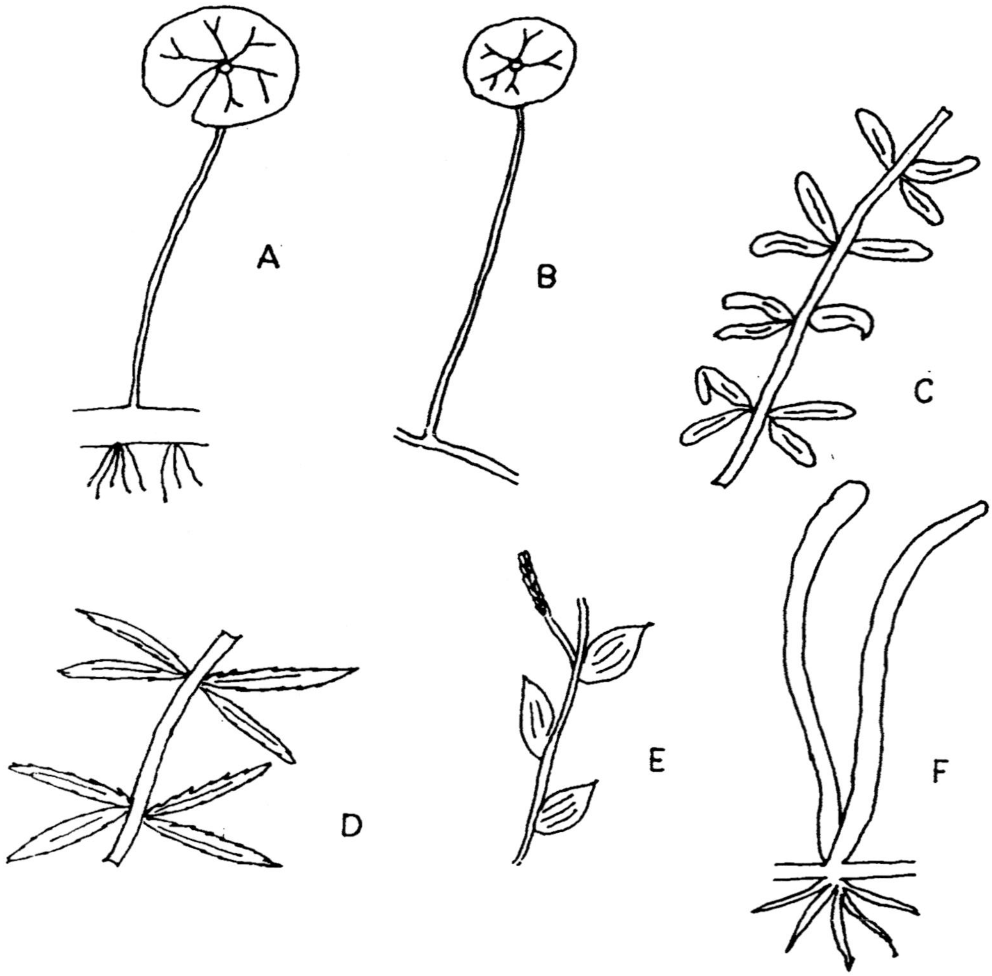

Fig. 3.3 : Some freshwater rooted higher plants with surface floating or submerged leaves.
A. *Nymphaea;* B. *Brasenia*; C. *Elodea*; D. *Hydrilla*; E. *Potamogeton*; F. *Vallisneria.*

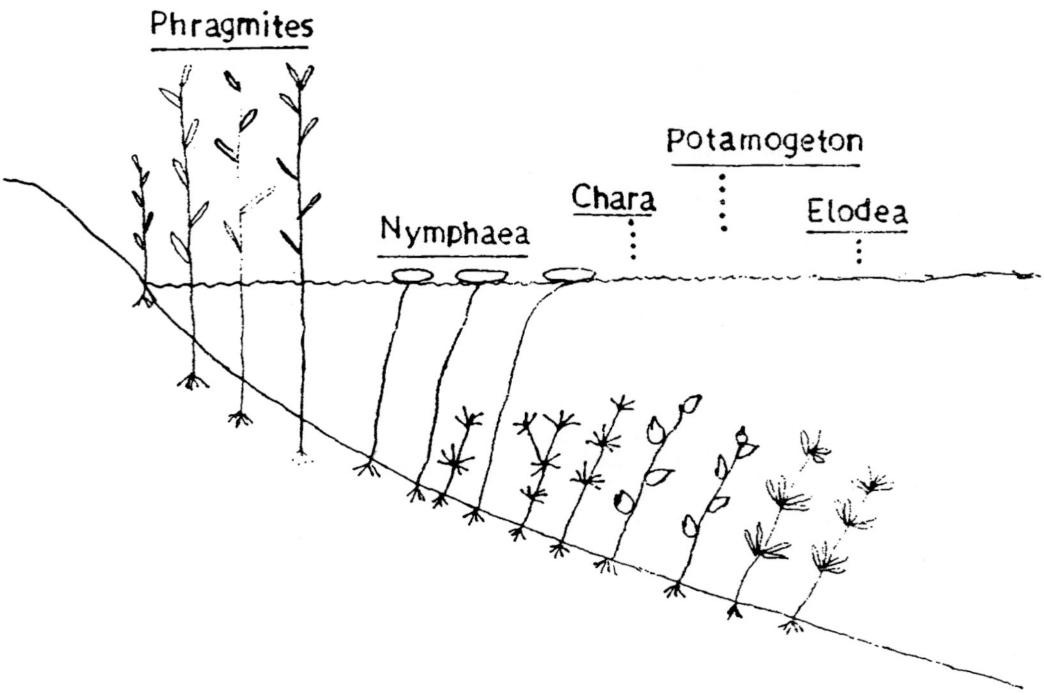

Fig. 3.4 : Section showing relation of plants to water environment.

IV

Lake Bacteria

4.1 OCCURRENCE

The bacteria occur in the lake either free-floating in the water column or on the bottom sediment.

(i) In the water column

The bacteria occur in a limited extent floating free in the water. The curve for the vertical distribution of bacteria follows very closely to that of phytoplankton (Fig. 4.1). Since, phytoplankton constitute the principal source of food for bacteria in the water column and also provide solid surface for their attachment, it is not surprising that the bacterial abundance is closely related to that of phytoplankton. The bacteria may also live upon the mucilaginous excretion products of the phytoplanktons as well as upon dead phytoplanktons. Apparently zooplanktons have fewer bacteria associated with them than phytoplanktons. Since dead and decomposing phytoplanktons also serve as a substratum or food for bacteria, the maximal bacterial count may occur after the phytoplankton start to die. In the water column, the bacteria are generally more abundant in the epilimnion than in the hypolimnion of fairly deep well stratified lakes, but the difference in the abundance of bacteria at different depth, disappears entirely during spring and autumn overturns and do not exist in the shallow lakes. The ice on the water surface generally contains few bacteria than the water just underlying the ice. It is possible that the nutrients may be somewhat more concentrated in the water adjacent to ice, because as water solidifies most of the dissolved solids remains in the liquid fraction.

Since bacteria multiply very slowly in the presence of low concentrations of organic nutrients present in oligotrophic lakes, the largest bacterial number usually occurs in eutrophic than in oligotrophic lakes, but whenever there is an influx of organic matter or an increase in the amount of suspended particles in the water to which the bacteria can attach (derived either by land drainage, stirring up of the bottom, the activity of photosynthetic organisms or fisheries activities in localized areas) their number can increase. The dystrophic lakes, which are rich in humus, occupy an intermediate position as far as the bacterial number is concerned and have different qualities of bacteria.

When the particulate substrates necessary for the favourable development of attached bacteria in the water column tend to settle to bottom, they carry with them the large number of bacteria from water column to bottom mud during sedimentation. Thus, the solid particles whose density exceeds to that of water will tend to reduce the bacterial population in the lake water. Similarly, the dead body or skeletal remains of planktonic organisms to which the bacteria are attached sink faster and faster towards the bottom and carry the attached bacteria along with them to the bottom.

The bacterivorous organisms and filter - feeding organisms like mollusks etc. which filter large quantities of water to obtain food are not a major factor in restricting the number of bacteria in the open water.

(ii) On the bottom sediments

The bottom sediments generally contains many more bacteria per unit volume then did lake waters (Fig. 4.1). The bacteria are associated with large quantities of dead organic matter accumulated on the bottom along with the solid substances. The bacterial population size of the bottom deposits appears to be more closely related to the organic content of the bottom than either to its distance from the land, to the depth of overlying water or to the composition of the lake bottom. The littoral benthic habitats occupied by large numbers of aquatic higher plants and algae and thus having large amounts of organic matter derived from these plants will naturally have more benthic bacteria than profundal benthic habitats containing little or no vegetation. The relatively large number of bacteria found on the clay bottom than on the silty or sandy bottom may be due to comparatively large surface area of the clay particles (which enhances the multiplication and survival of bacteria) then the silt or sand particles. One may find seasonal changes in the bacterial populations (both on the bottom deposits as well as in the overlying waters) in temperate lakes undergoing seasonal cycling of organic productivity.

4.2 CHARACTERISTICS

The bacteria usually associated with freshwaters are species of *Arthrobacter, Aerobacter, Flavobacterium, Achromobacter, Micrococcus, Pseudomonas, Sarcina, Spirillum, Vibrio, Bacillus* etc., but when the water is contaminated with sewage, the bacteria like *Escherichia coli, Clostridium, Proteus* and *Streptococcus* may also occur. The freshwater bacteria are usually smaller than the terrestrial bacteria, and most of them are motile rod shape, while *Cocci* are of limited occurrence (Table 4.1). Their number may range from less than 100 to several thousand per millilitre of water. The optimum temperatures for their growth are about 10 to 25°C and majority of them succumb when held for 10 minutes at 40 to 50°C. They multiply more rapidly at 20°C, but die more rapidly at higher temperatures. Nearly all of the freshwater bacteria are facultative aerobes and grow better in presence of oxygen then in its absence under laboratory conditions.

4.3 IMPORTANCE

The lake bacteria are very important as far as the decomposition of organic matter, mineralization of inorganic nutrients, oxidization of ammonia to nitrite and nitrate, transformation of sulphur compounds, liberation of phosphates, and other chemical reactions in the lake are concerned. The bacteria may themselves be ingested by certain animals as a source of food. The changes in the distribution of oxygen, nitrogen, carbon dioxide and other gases and in the hydrogen ion concentrations of the lake water, all are influenced by the microbial activity.

(i) Decomposition of organic matter

The chief function of the bacteria is the decomposition of virtually all kinds of organic matter into carbon dioxide, ammonia, certain minerals and water. It has been reported that around 60 percent of the organic matter assimilated by bacteria is oxidized and only 40 per cent is converted into bacterial protoplasm or intermediate products. The phycosphere recycling of essential inorganic nutrients by bacteria is highly advantageous under increasing conditions of nutrients limitation accompanying the proliferation of freshwater phytoplankton communities.

A large part of the plant tissues is ingested by grazing animals or is decomposed by bacteria and allied microorganisms. The grazing animals ingest a proportion of the primary producers, converting some of the organic carbon into animal's tissues, excreting varying quantities into the environment and oxidizing most of its as a source of energy. The grazing animals in turn may be ingested by predatory animals.

Juday (1942) estimated that the total weight of the plants is about six to twenty times that of the animals in inland lakes (Table 4.2). There is much more "dissolved" organic matter than "particulate" organic matter (weight of total live plants, animals and detritus) present in any lake. All plants and animals secrete and excrete and all organisms die and decompose and add some dissolved organic materials into the water. Of the particulate organic matter, upto 70 percent may be detrital and the remainder in the live plants and animals.

The so-called "dissolved" organic matter represents that part which is not removed with a high speed centrifuge. It also includes minute particles of decomposing organic material, some bacteria, colloids and organic matter in true solution. Although, some of the colloidal and minutely particulate organic matter can be assimilated by certain filter-feeding and mud-eating animals, most of the dissolved organic matter can be utilized only by bacteria and allied microorganisms. Juday (1942) pointed out that "the dissolved organic matter is in a constant state of flux", it is continually receiving both the excretory and decomposing materials from the biota on the one hand and losing organic substances that change over to inorganic compounds of bacteria and allied microorganisms on the other.

There is an extensive bacterial flora in the lake which is uniquely adapted to decompose chitin, lignin, waxes, cellulose, and other complex organic substances. The organic compounds which are most susceptible to be attack by bacteria are carbohydrates and simple proteins, lipids, lignins, chitin, polyuronides and other complexes known collectively as humus. The amount of oxygen consumed by bacteria in oxidizing organic matter in a given period of time is called the biological oxygen demand (B.O.D.). Thus B.O.D. is partly a measure of the quantity of organic matter and partly a measure of its biological oxidizability.

(a) *Carbohydrates decomposition* : The cellulose decomposing bacteria such as species of *Cytophaga* and *Cell Vibrio* are abundant on the bottom of lake. A part of the organic carbon is oxidized into carbon dioxide and a part is converted into bacterial protoplasm.

(b) *Lignin decomposition*: The lignin is oxidized less rapidly than other organic constituents occurring in the lake. Most of the species of *Micromonospora* found in the lake mud attack lignin.

(c) *Proteinaceous compounds decomposition* : Most of the lake bacteria are actively proteolytic. They rapidly decompose the simple and complex proteinaceous compounds with the liberation of ammonia and carbon dioxide.

(d) *Chitin decomposition* : The chitin is the chief constituent of the exoskeletion of crustacea, some mollusca, coelenterata etc. The chitinoclastic bacteria are most abundant on the mud surface and also in the intestinal content of the animals which feed on these animals. Most of the strains of *Micromonospora* isolated from lakes mud readily decompose chitin. The carbon dioxide and ammonia are the end products of chitin decomposition.

(e) *Lipid decomposition* : The fats are found to be far more resistant to be attack than proteins. Some of the lipolytic bacteria widely distributed in lake are *Pseudomonas, Sarcina* and *Bacillus* species. The fatty acids are decomposed into carbon dioxide and methane.

(ii) Gases in bottom deposits

Most of the bacteria consume oxygen and their oxygen consuming capacity is a function of the organic content of the bottom. In most bottom deposits, the oxygen consumption (due to bacterial respiration) is faster than its replacement either from atmospheric diffusion or from photosynthetic process. Although, the ratio of the mass of the standing crop of the bacteria to that of animals is about 1: 200, the bacteria produce carbon dioxide 30 to 15,000 times faster than animals per unit of biomass, indicating the relative importance of bacteria as carbon dioxide producer.

Under aerobic conditions, the organic matter is decomposed with the formation of carbon dioxide, phosphate and sulphate. The ammonia is liberated from nitrogenous compounds undergoing bacterial decomposition. The oxidation of ammonia into nitrite and nitrate occurs in shallow well-oxygenated lake bottoms where nitrifers analogous to soil *Nitrosomonas* and *Nitrobacter* occur.

While, under anaerobic conditions, appreciable quantities of methane, hydrogen and hydrogen sulphide are produced from the bacterial decomposition of organic matter. The methane is formed by the reduction of carbon dioxide in the presence of hydrogen during summer stagnation period by anaerobic bacterium *Pseudomonas methanica.*

$$4\,H_2 + CO_2 = CH_4 + 2H_2O$$

Appreciable quantities of hydrogen sulphide are also formed during anaerobic decomposition of proteins and reduction of sulphate by an obligate anaerobe *Desulfovibrio desulfuricans* occurring largely in stagnant bodies of water. This bacterium can utilize not only sulphate but also sulphite, thiosulphate and sulphur as hydrogen acceptor and the chief end product is primarily the hydrogen sulphide. On the other hand, the sulphur bacteria of *Thiobacillus* group (e.g. *T. thioparus*), widely distributed in coastal waters can oxidize hydrogen sulphide and other sulphides as well as sulphur, thiosulphate or tetrathionate into sulphate.

(iii) pH of water

Different lakes have different pH values. The bacterial activity has a pronounce effect on the hydrogen ion concentration of the lake water and mud. The following microbiologically activated processes will tend to increase the hydrogen ion concentrations of the water.

(a) production of carbon dioxide by respiring bacterial cells,

(b) Oxidation of hydrogen sulphide or sulphur to sulphuric acid or other acid sulphates,

(c) production of organic acids such as lactic, butyric, acetic and formic acids during decomposition of organic matters,

(d) assimilation of ammonia as a source of nitrogen or its oxidation as a source of energy,

(e) the formation of nitrite or nitrate, and

(f) the liberation of phosphate from the organic compounds.

However, the hydrogen ion concentrations of the water decreases and its pH increases by the following microbiological reactions.

(a) the utilization of carbon dioxide by either chemo-synthetic or photosynthetic autotrophs,

(b) Oxidation or decarboxylation of the salts of organic acids such as formate, acetate, propionate, lactate, etc,

(c) the reduction of sulphate to sulphur or hydrogen sulphide,

(d) the reduction of nitrate or nitrite, and

(e) the formation of ammonia from nitrogenous compounds containing amino acid, proteins, urea, purines bases, etc.

(iv) Utilization of hydrogen sulphide as hydrogen donor by photosynthetic bacteria

The photosynthetic sulphur bacteria such as *Chlorobium* and *Chromatium* (both obligate anaerobe) and *Rhodoseudomonas* (facultative aerobe) usually occur in shallow waters having sufficient light, but yet deep enough to have anaerobic conditions and large amount of organic matter (Fig. 4.2). In the presence of excess hydrogen sulphide (produced either by sulphate reduction or by anaerobic protein decay) and strong illumination, these bacteria can reduce carbon dioxide and deposit significant amounts of elemental sulphur and produce carbohydrate.

$$CO_2 + 2H_2S \xrightarrow{\text{light}} (CH_2O)_n + H_2O + 2S$$

The photosynthetic bacteria play a very minor role as primary producer in the lake.

(v) Bacteria as food for animals

Certain protozoans, ciliates, flagellates, snails, nematodes and mud-dwelling detritus feeders (detrivores) utilize bacteria (which are highly nutritious and readily digestible) as food. The bacteria play an important role in the food cycles by synthesizing cell substances and by converting waste or dissolved organic matter into particulate organic matter (bacterial cell substances) which can be utilized as food by animals. The ratio of dissolved to particulate organic matter is estimated to be more than 1 in different parts of the lake. The dissolved organic matter play no significant role in the direct nutrition of aquatic animals, but the bacteria serve as an intermediary agent of vast importance, mineralizing roughly 70 percent of it and converting 30 percent of it into bacterial cell substances or intermediate products.

(vi) Reciprocal relations of bacteria and plants

Bacteria play an important role in the production of inorganic nutrients including ammonia, nitrate, nitrite, sulphate, phosphate, as well as large quantities of carbon dioxide, which are essential for plants growth. The composite bacterial population of lake may produce more carbon dioxide than the combined animal population. While plants are the chief source of organic matter for saprophytic bacteria. The plants are attacked by bacteria immediately after their death. The secretion of organic compounds by living plants can also provide food to associated bacteria.

(vii) Reciprocal relations of bacteria and animals

The digestive tracts of most water animals contain bacteria which are very versatile biochemically, being capable of attacking a wide range of materials. Collectively, these bacteria secretes enzymes which attack many of the substances swallowed by animals for which they themselves have no enzyme. The power to break down the pectins, hemicellulose, cellulose, lignin, chitin and other organic complexes is much more widespread among enteric bacteria than among animals which ingest these organic complexes. While, secretions of animals are a source of food for enteric bacteria.

(viii) Bacteria as algal pathogens

(See chapter 7, section 7.2)

(ix) Bacteria as animals pathogens

Bacillus columnaris infect the epidermis, gills and fins of several species of freshwater fishes. Any diseased fish which can no longer swim as fast as his companions may soon be captured and eaten by a larger fish or other predator and consequently the pathogen which incapacitates the host is also destroved.

Table 4.1 : Comparison of main morphological groups of bacteria in lake water.

Morphological group	*(%)*
Gram - negative rods	95.5
Gram - positive rods	3.8
Cocci	0.7

Table 4.2 : Summer standing crop in Wisconsin lakes given as kilogram per hectare of lake surface on a wet-weight, ash-free basis (from Juday, 1942).

	Lake Mendota	*Green lake*	*Nebish lake*	*Weber lake*
Phytoplankton	1875	2767	608	1069
Bottom flora	4600	4218	590	553
Total weight of plants	6475	6985	1198	1622
Zooplankton	120	177	42	74
Bottom fauna	414	138	122	147
Fish	-	-	35	23
Total weight of animals	534*	315*	199	244
Dissolved organic matter	15201	27901	3829	2866

*Excluding fish, the standing crop of which is not given.

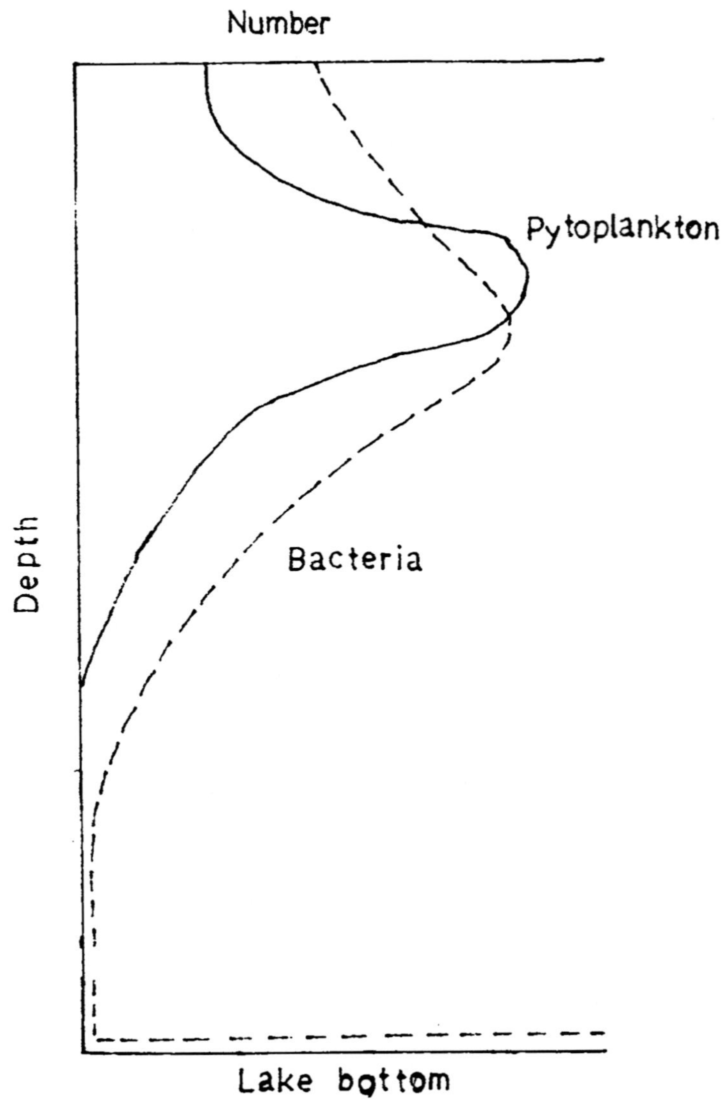

Fig. 4.1 : Vertical distribution of phytoplankton and bacteria in a lake.

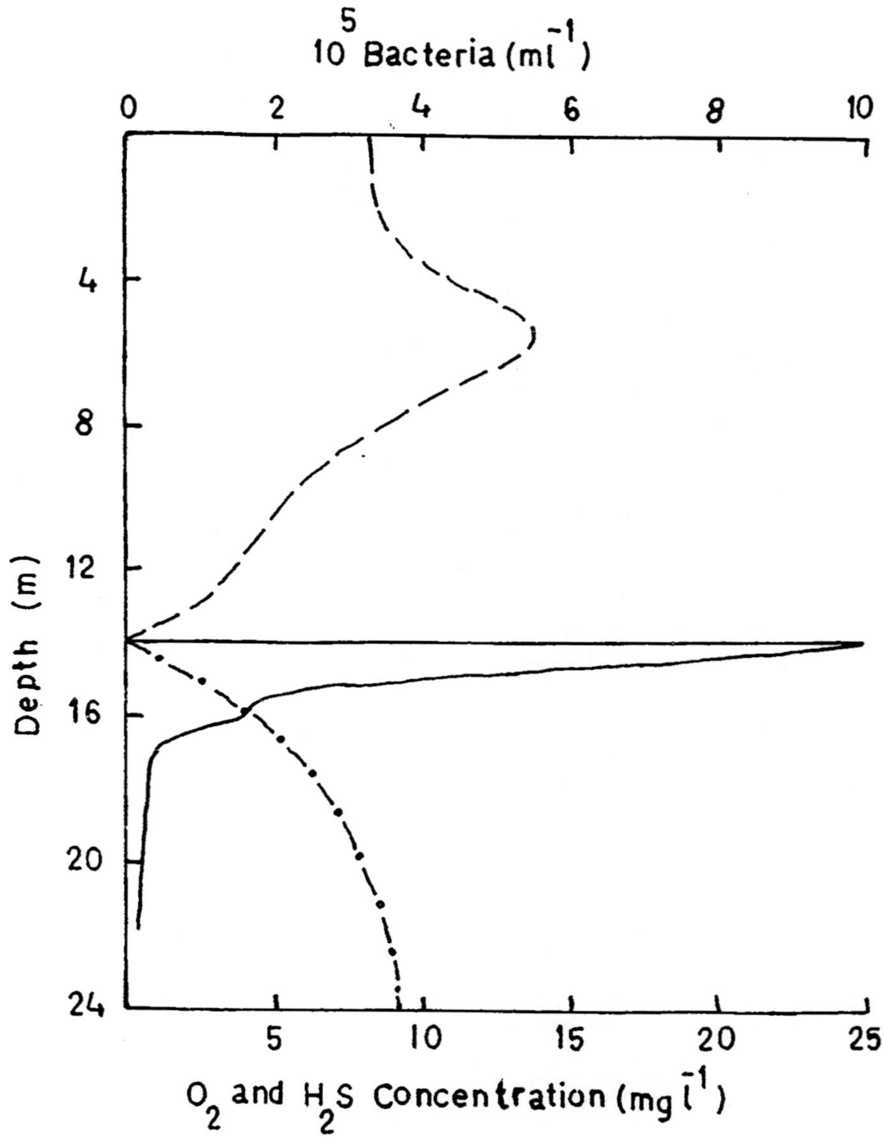

Fig. 4.2 : The distribution of *Chromatium* sp. (———) in relation to depth, and oxygen (– – – – – –) and hydrogen sulphide (- - - - - -) concentration in a lake rich in organic matter.

V

PHYSICAL FACTORS AFFECTING THE FLORA

5.1 LIGHT

The most important physical factor affecting an algal flora is the light, because of its necessity for photosynthesis. The light generally act by way of photosynthesis which produce carbon skeletions and which acts as sink for incorporated ions, provides energy necessary for active transport and production of charged polymers and stimulates protein synthesis and growth, and which results into increased nutrients uptake. The light also controls the diurnal migration and vertical distribution of phytoplanktons. The solar energy generates wind which coupled with surface layers of water can produce water currents.

5.1.1 Light compensation intensity and depth

The light compensation intensity is that light intensity at which plants respiration equalizes the photosynthesis (this may fluctuate widely from 400-1000 lux for different phytoplankton species) and the water depth at which this light intensity (usually considered to be about 1 percent of the incident light intensity at the water surface) occurs is called the light compensation depth.

There is not a linear relationship between the radiation flux and the rate of photosynthesis. A feature which is quite apparent when summer and winter carbon fixation rates are compared. The winter rates of photosynthesis are nowhere near as reduced as might be expected from the reduction in irradiance. At low temperatures, the light compensation intensities also decreases and this might be due to decrease in respiration rates at low temperatures. For example, the light compensation points for certain algae at 16°C may be 350 to 400 lux, but at 10°C it get reduced to 250 to 300 lux. Thus, the low light intensities will be sufficient to optimize the photosynthesis at low temperatures and this might suggest that in winter season even with low light intensities, photosynthesis may be high.

A common way to measure the relative transparency of water is to lower in water column, a *Secchi disc*, which is a white plate of 20 centimetres in diameter attached to a cord marked off in linear units, and mark the depth at which the disc disappear from the sight. The disc is lowered a bit farther, then raised until it reappears and that depth is marked. The two depths are averaged. The light intensity at the depth of disappearance of the disc is usually considered to be about 5 percent of that at the water surface. The other more exact procedures employed to measure the relative transparency of water includes photographic methods, pyrilimnometers or photoelectric cells.

5.1.2 Fate of incipient radiation

The depth of light penetration into water is affected by several factors such as the intensity of incipient light, the angle of ray incidence, reflection at the water surface and scattering and absorption within the water. Thus, the light compensation depth of an alga varies not only with the latitudes but also with the time of day, season and water clarity. The penetration of light anywhere in the water is reduced either when the sun is away from the zenith, at high latitudes or in winter season. When the light strike the water surface, a certain amount of light is reflected back into air depending upon the angle at which the light strike the water surface. If the striking angle from the horizontal water surface is low, a large amount of light will be reflected. Conversely, nearer the angle is to 90 degree (that is perpendicular to the horizontal surface of the water), the greater will be the penetration of light into water and lesser will be its reflection (Fig. 5.1). In tropical regions of earth, the sun is directly overhead at midday and this position changes little with seasons of the year, and so the light conditions are maximal all the year long, but as one progresses towards north or south from the equator, the sun may be directly overhead only during summer months, but may be far from this position at other times of the year and as a result, the amount of photosynthesis also changes, being reasonably high in the summer and much lower in the winter. Near the poles, there is almost no solar radiation during winter months, but the amount of light received in summer will actually exceed that at the equator. The factors such as clouds and dust also act to interfere with light in such a way that lesser amounts of light passes through atmosphere. This reduces the availability of light initially without any reference to water conditions (Fig. 5.1).

The portion of light that enters the water column is subject to further reduction from two additional process acting on it within the water. The first is the absorption of light by water itself, by solutes or by suspended materials both inanimate and animate, and the second is the reflection from various suspended particles in the water column. The suspended living or dead particles intercept the light and may either absorb it or reflect it back into the water, where a portion may escape into the atmosphere and a portion may again be reflected back into the water (Fig. 5.1). The absorption of light by water molecules is a function of wavelengths.

The light is not only rapidly but selectively absorbed. In pure clean water, the red light is absorbed most rapidly at a rate of 64.5 percent per metre, orange at 23.5 percent per metre, yellow at 3.9 percent, green at 1.1 percent and blue at only 0.52 percent per metre. The blue light penetrates the farthest. The violet light is absorbed at 1.63 percent per metre Very little ultraviolet light penetrates the water and nearly all infra-red wavelengths are absorbed in the first metre. The suspended matters and dissolved substances in the water absorb most efficiently the blue portion of the spectrum and shift the maximum transmission to green light or even to red light. The dissolved yellow substances, which are a part of the pigment–protein complex in the intact cells, are rapidly released from the algal cells into water and can compete with the algae for the absorption of blue light. The chlorophylls absorb most efficiently in the violet-blue and orange-red portions of the spectrum, and least efficiently in the green-yellow portion. Therefore, deeper zones of transparent lakes may not have those wavelengths of light that are most efficient for photosynthesis of algae and higher plants.

When the large amounts of suspended materials are present in the water, the depth of light penetration in the water is severely reduced. The light penetration is also affected by the abundance of both phytoplanktons and zooplanktons in the water. As the phytoplankton population grows, they themselves absorbs more and more light and as a result the light compensation depth begins to move upward and will decrease (Fig. 5.2).

The apparent colour of lake water depends upon the sky reflections, the colour of the suspended materials, the colour of the chemical contents dissolved in the water or that of plants, animals or bottom of lake. The blue colour of pure water is as a result of blue light scattering by water molecules. A green colour is usually associated with high concentrations of calcium carbonate in the water, while waters from bogs or swamps contain humic materials and are often dark brown in colour.

The light intensity decreases exponentially with the depth of water (Fig. 5.3). In very turbid lake, 99 percent of the incident light is absorbed at a depth less than 2 metres, while in a lake of medium turbidity, the same about is absorbed by a column of water 10 metres deep, but in a clearest lake it is not absorbed even at a depth of 20 metres. This difference is obviously a great enough to be a major factor in controlling the survival of algae in a water body. In general, the light attenuation with water depth follows a straight line on a semilog plot and a curve line on a linear plot (Fig. 5.3).

5.1.3 Energy flow through organisms

The algae are usually more suitable as a food for much wider variety of organisms than most of aquatic higher plants, because they replenish quickly, have short life cycle, less turn over time, decompose quickly and completely and have higher nutritional values than higher plants.

The energy entering into water diffuses through a complex web of physico-chemical processes and from one organism to another with the loss in the form of heat at each stage and ultimately a residual fraction of the initial energy input is deposited on the sediment to contibute to the store of fossil fuel. The flow is thus a one way system which is balanced by the influx of solar energy. The total incoming energy at the water surface is partially lost back to the atmosphere through reflection, a part is absorbed as heat and a part is absorbed by plants and utilized in photosynthesis. The conversion of light energy into chemical energy by plants is only 1 to 5 percent of the absorbed energy, and the energy is then transmitted to herbivores. The algal materials eaten by herbivores then passes into carnivores and is further deposited on the sediment as detritus, which is either the dead parts of the plants, animals or the unconsumed particulate organic matter or faeces. A small part of energy also exist in the water in the form of dissolved organic materials secreted by plants and animals. It is well known that lakes with a luxuriant flora of all kinds have a correspondingly dense population of animals. In addition to serving as food for the microfauna, either directly or indirectly, phytoplanktons makes up a large part of the food consumed by several species of fishes and certain mollusks. Snapping turtles eat *Spirogyra crassa*, *Cladophora* and *Ceratophyllum* species. The detritus from all plants and animals sources is incorporated into sediment. A small part of it is consumed by detritivores and ultimately most of it is decomposed by bacteria and fungi. The bacteria and allied microorganisms can also take up the dissolved organic materials present in the water to partly fulfil their energy requirement. The gross efficiency at each step in the food chain is often quoted to be 10 percent.

5.1.4 Irradiance and algal photosynthesis

The algal photosynthesis is light dependent. At low irradiance levels, the photosynthetic rate is limited by the light reactions of the photosynthesis, involving low numbers of available light quanta. The rate of photosynthesis increases linearly with an increase in irradiance until a plateau is reached at some intermediate irradiance levels and a further increase in irradiance has no effect on photosynthesis though very high irradiances are inhibitory (Fig. 5.4). On the other hand, the rate of respiration of algae does not change much with the light intensity (photorespiration may occur in presence of light). This means that as we go deeper in the water column, the rate of algal photosynthesis declines, because of decrease in light intensities, and at some depth the rate of photosynthesis will equalize the respiration. No alga or plant can persist for other than short periods of time below its light compensation depth.

The different algal species have different light optima and saturating light intensities for photosynthesis. For example, the green algae have been observed to have light saturation values of about 500-750 foot candle (ft-C) for photosynthesis, the diatoms of about 1000-2000 ft-C and dinoflagellates, of about 2500-3000 ft-C (Ryther, 1956). These values are obviously only approximation and the exact values will depend upon many factors including 'sun' and 'shade' adaptations of an alga. The production versus light intensity curve of benthic diatoms has a broad optimum than either the planktonic dinoflagellates or green algae (Fig. 5.4).

5.1.5 Depth-Photosynthesis profile

It is generally agreed that the light intensity is the limiting factor in determining the maximum depth at which an alga can occur in a given body of water. Near the water surface, where the light intensities are highest, most species shows a leveling off or a decrease in photosynthesis and this might be either due to saturation of photosynthetic apparatus in such a way that it is not possible to increase the photosynthetic rate further at high light levels or due to photooxidation of enzymatic reactions. The surface inhibition has been shown to be more pronounce around mid-day in different lakes, suggesting that it is a function of light intensity. The maximum algal production occurs at a depth where the light intensities are about 40-50 percent of light intensities at the water surface (Fig. 5.5). Thus the peak of algal production may be at a greater depth in sunny or summer days than in cloudy or winter days. The shape of the production curve (carbon fixation curve) in the water column does not depend only upon the time or type of day or season but also upon the richness or turbidity of the lake and also upon the preferred place of occurrence of algae in the water column. For example, in a eutrophic turbid lake, the carbon fixation curve has its maximum in the upper epilimnion and in a clean water mountain lake only a deep maximum is observed (Fig. 5.6 a and b, respectively). Some blue-green algae prefer residence in metalimnetic zone and then an upper and lower maxima of carbon fixation may occur (Fig. 5.6 c) or there may not be any distinct maximum occurrence (Fig. 5.6 d).

In most lake waters, nanoplanktons are present more in number than net planktons. They have high surface area, high chlorophyll content, rapid turnover rates and are thus responsible for more carbon fixation than net planktons. In some lakes, over 90 percent of the total carbon fixation during favourable season may be due to nanoplanktons (Fig. 5.7).

There is rarely any exact correlation between chlorophyll content and carbon fixation. Both chlorophyll content and carbon fixation of phytoplankton varies with time. In all cases, the maximum carbon fixation occurs higher in the water column than the chlorophyll maximum (Fig. 5.7). The chlorophyll amount increases much more than cell number in deep waters, indicating the shade adaptation. The ratio of chlorophyll content to carbon content varies by a factor of nine or more in different algae at different depth.

5.1.6 Primary production and algal biomass

The biomass measurement is not necessarily an index of rate of production in an alga. An alga with large biomass may contribute very little to carbon fixation than the one with small biomass. In general, the smaller cells having low carbon content and short renewal times are metabolically more active than larger cells having more carbon content. Thus, as has been long suspected, a species comprising only a small proportion of the total algal biomass may make an inordinately large contribution to the carbon fixation. Conversely, organisms present as large components of the total biomass can contribute very little to carbon fixation. For example, in Castle lake, at 1 metre depth, *Cyclotella* and *Microcystis* comprised the largest amount of biomass, while the most productive alga was a *Peridinium* sp which contributed 32 percent of the community primary productivity, but has only 1 percent of the total algal biomass (Stull et.al., 1973). The curves of primary productivity and algal biomass commonly shows an inverse relationship (Fig. 5.8).

In large deep lakes, the proportion of primary production contributed by benthos is always much smaller than by phytoplanktons, but in a lake Michigan dune pond, 26 percent of the total fixation was due to epipelon, only 13 percent due to phytoplankton and the remainder by macrophytes.

5.1.7 Adaptations in algae in response to light

(i) Sun-shade adaptation

There is no doubt that sun-shade adaptation does exist in natural conditions. It has been observed that the optimum light intensities as well as the light compensation intensities both are much lower in deep growing shade-adapted algae then in sun-adapted surface algae (Fig. 5.9).

With decreasing irradiance at increasing water depth, a two to ten time increase in concentrations of photosynthetic pigments associated with an increase in number of thylakoids, size of chromatophores and number of chromatophores may occur in algal cells. The shade adapted cells are thin walled, more efficient in absorbing dim light and are able to grow at low temperatures. In them, the photosynthesis achieve saturation at low levels of light intensities and the chromatophores may move towards cell walls facing the light in order to absorb the dim light efficiently.

However, the sun-adapted cells are less sensitive to photoinhibition at high light intensities. They often have an increase in carotenoids to chlorophylls ratio, which may protect them against intense light. The intense light may cause the photochemical breakdown of chlorophylls. The members of blue-green algae and of some green algae are protected against intense light by having thick mucilaginous sheath. The chromatophores of some algae may move towards side walls to avoid high irradiation damage.

Since, there is a randomized circulation of phytoplankton cells in the water column, no phytoplankton cell experience anything like a relatively constant light environment, and if the water movement is rapid, one can perhaps assume that the phytoplankton cells will not have time to acclimatise to either high or low light intensities, as they move through the water column and therefore 'sun' and 'shade' adaptations do not exist in the phytoplanktons, but they exist certainly in the benthic algae.

(ii) Chromatic adaptation

Engelmann (1882) suggested that in an alga, the ratio of different pigments is adjusted according to the wavelengths of light to which it is exposed. Since, blue-green wavelengths dominates the low light environments of the deep water, in deep water algae a shift in the composition of pigments towards those absorbing maximally the blue-green light will occur (e.g. chlorophyll a + c were observed to increase in blue light and carotenoids in green light). A blue-green alga when grown in green light has higher relative proportion of red phycoerythrin, and when grown in red light has higher relative proportion of blue phycocyanin. Thus, the colour of an alga will change according to the colour of illumination. The trichomes of *Oscillatoria sancta* will be green in colour in red light, blue in colour in green light, and yellow in colour in blue-green light. These changes are due to changes in relative concentrations of different pigments present in an alga.

(iii) Heterotrophic growth

In polar lakes, the light is very much limited due to long period of darkness in water and it has been suggested that some heterotrophic growth may occur in algae. Various organic substrates such as glucose, acetate, amino acids, peptone and proteins present in trace quantities in natural waters may allow some members of Volvocales, Crytophyceae, Euglenophyceae and Bacillariophyceae to grow heterotrophically on these organic substances. The heterotrophy may help the phytoplanktons to supplement their low rate of autotrophic carbon assimilation near the compensation depth, and there is a possibility for the occurrence of healthy algal cells far below the euphotic zone due to heterotrophy. An alga like *Ochromonas malhamensis* possess pigments in such small amounts that this alone is not sufficient for its growth and the alga must be supplied with acetate, palmitate or glucose for better growth.

5.2 TEMPERATURE

The temperature is a measure of energy of molecular motion. The range of temperature variation is smaller and occur more slowly in the water than in the air. The annual range of temperature variation is greater in the median latitudes (temperate regions) than in the lower (tropical and subtropical regions) and higher latitudes (arctic and antarctic regions).

5.2.1 Optimum temperature for algal growth

The different groups of algae may have different temperature optima for the growth. For example, the maximum temperature for chrysophytes growth tend to be about 24-27°C and this in part explain their presence in arctic lakes and rare occurrence in warm water lakes. The optimum temperatures for the growth of most diatoms are found to be around 20°C. The diatoms *Nitzschia frigida, Amphiprora hyperborea* and *Cylindrotheca closterium* are common under-ice flora during winter stratification in temperate lakes. Most of the green algae favour the temperatures of 20-30°C for their growth and when the temperatures are raised above 35°C, the usual flora of diatoms and green algae are replaced by excessive growth of few blue-green algae. The blue-green algae which are able to tolerate excessive temperatures of 40-85°C are *Synechococcus lividus, S. elongatus, Lyngbya contorta, Mastigocladus laminosus, Oscillatoria filiformis, Phormidium laminosus, P. tenue, Plectonema natatum, Scytonema coactile, Pleurocapsa fluviatilis and Cyanidium caldarium.* Therefore, the temperature of water may decide that which group of algae is more favoured in that water. But sometimes an algal species may exist and grow in both temperate and tropical lakes, and this may be due to occurrence of cold and warm water races of that species.

5.2.2 Influence of temperature

The temperature of water can influence the growth of algae both directly and indirectly.

Direct effects

(a) *Algal metabolism* : An increase in the biological activities of algal cells like rates of diffusion, carrier mediated uptake, cell division and overall growth by a factor of about 2 for each 10°C rise in temperature upto certain optimum temperatures and thereafter their decline with further increase in temperatures has been observed.

Certain algal species have narrow range of temperature tolerance for their growth and are called *Stenothermal,* while others are able to withstand and grow in wide range of temperatures and are called *Eurythermal.* In general, the diversity of algae in any water body increases as one approaches towards the optimum range of temperatures within the range of tolerance for most of the species of a community, but when the temperature regime moved away from the optimum for most of the algal species either by becoming colder or hotter, the diversity of algae decreases.

(b) *Change in light compensation point* : A decrease in respiration rates with decreasing temperatures is expected to lower the light compensation point, and an increase in respiration rates at higher temperatures will increase the light compensation point. The lethal effects of high temperatures is due to an increase in respiration rates, fast depletion of reserve food and protein denaturation. Since at low temperatures, the light compensation point decreases, the phytoplanktons of colder lakes are capable of surviving long periods of darkness and shows high net photosynthesis, but if the temperature is raised, the dark tolerance time is greatly reduced. Therefore, in winter season, an alga will have lower light compensation point and lower light optima than in summer season. At very low levels of illumination, it would not be possible for an alga to survive unless the temperature is also not lower. The shade-adapted alga usually grow at lower temperatures than sun-adapted alga.

Indirect effects

(a) *Solubility of oxygen in water* : The solubility of oxygen in the water decreases as the temperature of water increases (Table 5.1). The high temperatures will further aggravate the dissolved oxygen problem in water by increasing the rates of plants, animals and bacterial respiration and by favouring the decomposition of organic matter. The deficiency of dissolved oxygen in the water at high temperatures may prove limiting to the growth of plants and animals.

(b) *Viscosity of water* : An increase in temperatures will lower the viscosity of water and thus increases the settling speed of suspended particles and sinking rates of planktonic algae. It is well known that diatoms living in warm water often has less silica in their cell walls than those living in very cold water.

(c) *Convection currents in water*: The surface water heating at day time and its cooling in night may cause slow convection currents in the water. The heat can also be stored in the bottom water and can be an important source of energy for slow convection currents particularly in ice covered lakes. These currents may be important in preventing the sedimentation of algae and to maintain them in suspension.

(d) *Summer stratification* : The sunlight contains infra-red wavelengths and when absorbed by water produce heating effect. The magnitude of heating depends upon the size of water body and its ambient temperature. The continued surface heating of water decreases its density and results into horizontal stratification of water into two or more layers. The warm, circulating, upper surface layer is called epilimnion; while the deeper, more viscous, non-circulating cooler layer as hypolimnion. A thin intermediate layer called thermocline or metalimnion may exist between epilimnion and hypolimnion and in which the temperature of water decreases abruptly, generally involving a drop of at least 1°C per metre of depth and occasionally as much as 7°C per metre, and a subsequent density differences between epilimnion and hypolimnion prevent the exchange of water between them (Fig. 5.10). As the summer progresses, the thermocline sink lower and lower and the volume of epilimnion increases and that of hypolimnion decreases. The thermocline generally determines the depth of phytoplankton circulation in the epilimnion. Those phytoplanktons which are trapped below the thermocline tend to sink and will die unless they swim sufficiently to reach the upper layers or the thermocline break down. The temperature of hypolimnion declines gradually from the lower edge of the thermocline to its bottom (where it is seldom below 4°C), but that of epilimnion remains nearly uniform (at above 20°C) down to upper edge of the thermocline due to thorough mixing of water.

In large, deep tropical lakes, because of the large volume of water, there is only slight diurnal, seasonal or annual variations of water temperatures and summer stratification is a year-round phenomenon; but in small shallow tropical lakes, because of their smaller volume of the water, the diurnal temperature fluctuations can be more than 10°C, and the summer stratification which develops during day time can break down at night. In polar lakes, the surface water will remain cold throughout the year and very little or no thermocline develops here. But in temperate lakes, the seasonal fluctuations of temperature is more pronounce with warm summer and cold winter, resulting into subsequent formation of summer and winter stratification.

The summer stratification has pronounce ecological implications, because the water of epilimnion is usually nutrients-depleted and that of hypolimnion is nutrients-enriched and is mostly out of the reach of the algal cells usually present in the epilimnion. If the thermocline lies below the euphotic depth, as is usually the case, the oxygen supply is cut off in the hypolimnion. The extent of depletion of oxygen in the hypolimnion also depends upon the amounts of decaying organic matter present there. The productively rich lakes are generally subject to greater oxygen depletion during summer than the poor lakes. But, on the other hand, if the water of lake is very transparent and the euphotic depth is more than the depth of thermocline, the phytoplanktons will growth even below the thermocline in the upper part of the hypolimnion and the oxygen may be present in greater abundance in hypolimnion, since cold water holds more oxygen than warm surface waters.

(e) *Winter stratification* : With the onset of winter in temperate latitudes, the surface water cool down to zero degree centigrade. It expand, becomes lighter and remains on the surface and freezes. Immediately below the ice, the temperature of water is very close to 0°C, but in 1 or 2 metres of additional depth it usually rises rapidly to about 4°C (Fig. 5.10). In the winter, the oxygen content of the water is not greatly reduced, because cold water holds more oxygen and the rates of organic matter decomposition and respiration are not so great. An appreciable amount of light penetrates the ice and allow phytoplanktons to carry out photosynthesis just below the ice. The winter stratification is not as severe to plants and animals as summer stratification, but an exception may occur when heavy snow cover the ice and cut off the light penetration into water (from 20 percent during the ice cover to 2 percent when snow cover the ice), resulting into decrease in algal photosynthesis, oxygen depletion in water and suffocation of fishes.

(f) *Spring and autumn overturn in temperate lakes* : As the ice melt during onset of spring and the surface water warms up to 4°C, the water of entire lake will be at the same temperature and undergo vertical turnover (Fig. 5.10). This situation may last for several weeks depending upon the weather conditions.

Similarly, with the onset of autumn following summer, the temperature of epilimnion drops until it becomes more or less equal to that of hypolimnion, and the water of entire lake begins circulating and the oxygen again returned to the depths. The water is then uniform in temperature and density at all depth and even a slight wind on the water surface can produce complete water circulation (Fig. 5.10). The autumn overturn last for several weeks until ice formation starts on the water surface in winter.

5.2.3 Extreme temperatures tolerance in algae

(i) Low temperature tolerance

The arctic and antarctic algae are partially protected from starving under ice by having a progressively decreased metabolism as the winter goes on. As the temperature of water is lowered, the rate of respiration decreases more rapidly than that of photosynthesis. Many algae occur in encysted or zygospore stage during winter months of freezing. The charophytes may overwinter beneath the ice in the form of storage bulbils or secondary protonema. They also reestablish annually by germination of oospores, which appears to require a period of rest, low temperatures and low oxygen levels to initiate germination.

(ii) High temperature tolerance

The mechanism of heat hardiness is not well understood, but it is thought to be associated with resistance to enzyme denaturation. The presence of thick mucilaginous sheath, high osmotic potential of cytoplasm, lack of vacuoles and formation of akinetes in blue-green algae may help them to tolerate high temperature exposure.

5.3 WAVES AND CURRENTS

No water mass is stationary. Even the water of a smallest pond is in constant oscillation or movement produced primarily by the action of wind. The wind is caused by differential heating of the atmospheric air masses aided by the Coriolis force (the result of rotation of earth on its axis). The effect of wind action on the water surface depends largely upon the extent of the exposed water surface, the presence or absence of protecting upland and the configuration of lake relative to the prevailing wind direction. In addition to wind, the water movement is generated also by inflow-outflow movements, changes in barometric pressure, diurnal heating and cooling, earthquakes, volcanic explosions and underwater landslides.

5.3.1 Waves

The waves are surface water oscillations in which no horizontal transport of any individual water molecule occur and the water molecules transcribe a circle, moving upward and to one side as a wave crest approaches and to other side and downward near to their original position as the wave crest passes by. The waves once generated by wind action are transported horizontally outward and away from the centre of origin. The depth of waves action depends largely upon the strength of prevailing wind and it may be sizable in large lakes. As waves enters the shallow water of shores and begins to encounter the frictional resistance of shore bottom, they slow their forward motion and their wavelength decreases (Fig. 5.11). As a result, they begin to increase in height and become steeper. At a point where water depth is 1.3 times the height of waves, the waves will break, releasing the energy on the shore.

5.3.2 Currents

The currents are water movements where the winds are steady and persistent in direction and that results into slow horizontal transport of water molecules or masses across distances. The movement of water masses need not necessary to be only horizontal on the water surface, but may also be vertical involving the whole water depth, e.g. spring and autumn overturns in temperate lakes.

Water currents in different seasons in temperate lakes

***(i)* During spring and autumn**

Since there is no density differences of water during spring and autumn, the winds blowing across the water surface creates currents that are forced all the way to the bottom of lake by the shape of lake basin and the entire volume of lake water is mixed thoroughly using the energy of wind.

The surface water which is slightly warmer by the influence of sun and warm air is driven down by the action of wind and is mixed with the deeper, cooler water (Fig. 5.10), so that a uniform water temperature is maintained from top to bottom and in this way some heat is carried down in the lake at a much faster rate than it would has otherwise be possible only by thermal conductivity. The turbulent mixing of water will take phytoplanktons intermittently below the photic zone and many reduce their bloom.

(ii) During summer

As the summer progresses, the surface water of lake gets warmer and warmer and the density differences between the surface water (epilimnion) and the deeper water (hypolimnion) becomes greater and greater, so that even strong winds are no longer able to drive the surface currents all the way to the bottom because they can not penetrate the dense water below, instead the surface currents of epilimnion returns along the top of the denser layer of thermocline, that serve as a barrier between epilimnion and hypolimnion (Fig. 5.10). Below the thermocline in deep water of hypolimnion, a slow convective current may move in the direction opposite to that in the epilimnion.

Such condition may exist throughout year in deep tropical lakes.

(iii) During winter

In temperate lakes, during winter, the water freezes from the top and the ice forms over the heavier bottom water. The bottom water is cut off from the wind stress and the circulation of lake water will be slow down until the ice melt in the spring. Due to storage of heat in deep bottom water, a slight convective current may occur in the underlying water, which will help algae to be in suspension and prevent their complete deposition on the bottom sediment.

Influences of waves and currents on algae

(i) Algal metabolism

The relative movement between the algae and water results into exposure of algae to fresh media and thus increases the metabolic activities of algae by creating steeper diffusion gradients next to cells. A number of studies have shown that the rates of nutrients uptake, respiration, photosynthesis, growth and removal of extracellular products, all are higher in algae growing in moderate currents than in still water. An algal species occurring in stream has higher metabolic rates than in lake water.

However, high water turbulence may be inhibitory to growth and cell division in dinoflagellates. There is an observed tendency of dinoflagellates to avoid the top 0.5 metre of the water column and this is attributed primarily to the wind induced water turbulence and secondarily to the presence of excess light at the water surface. Some dinoflagellates are very delicate and tend to be immobilized when agitated. They prefer stable waters and can not tolerate mechanical shearing of high turbulence.

The water current or wave action can also increase the turbidity of water and can cover the benthic algae with muddy sediment and can deprive them of light. The rough wavy water surface reflects much of the incipient light back into atmosphere than the smooth water surface.

(ii) Position of algae in water column

The water turbulence is responsible to maintain most of the unflagellated phytoplanktons suspended in the water column and keep them near the water surface or subsurface for longer time periods for optimum photosynthesis and also prevent their permanent loss by sedimentation. The water turbulence increases the dissolved oxygen content of the water by helping surface dissolution of oxygen from the atmosphere. However, disturbances by turbidity currents can completely mix the fossil algal assemblage and can destroy their sequential pattern of deposition on the lake bottom.

The phytoplanktons are transported horizontally from one region of the lake to other by surface water currents and in their way they encounter various changes in physical and chemical factors which will consequently produce some changes in the composition of moving flora. The water currents may also remove the loosely attached metaphyton from its host surface.

The free-floating higher plants like *Lemna, Wolffia* etc, which do not form any interconnected mats are restricted to narrow areas along lake margin where water currents are negligible.

The downwelling water currents can carry phytoplanktons in waters deeper than their light compensation depths and will thus influence their photosynthesis adversely, but will expose them to nutrients enriched bottom waters, and if these cells are brought back to the surface water by upwelling water currents, they will serve as a fresh inoculum for algal growth in upper waters. The upwelling water currents will also provide replenishment of nutrients to upper water.

Table 5.1 The solubility of oxygen in freshwater at different temperatures.

Temperature (°C)	*Dissolved O_2 (c.c./L)*
0	10.29
5	9.03
10	8.02
15	7.22
20	6.57
25	6.04
30	5.57

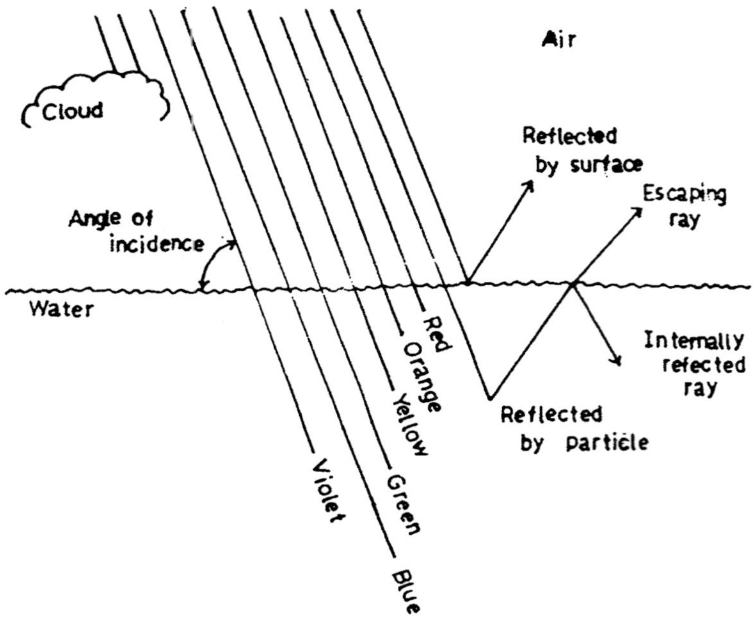

Fig. 5.1: The fate of incipient light in the clean water lake.

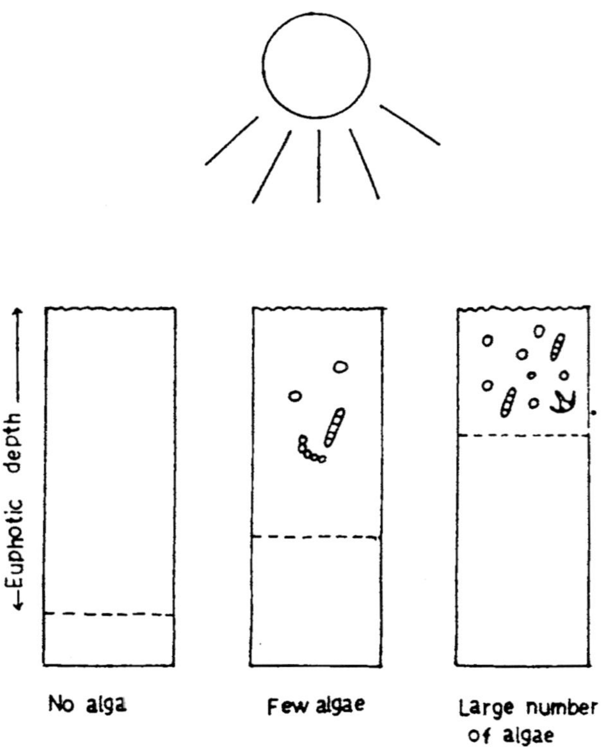

Fig. 5.2: Changes of compensation depth with increasing algal numbers. Dashed line indicates compensation depth.

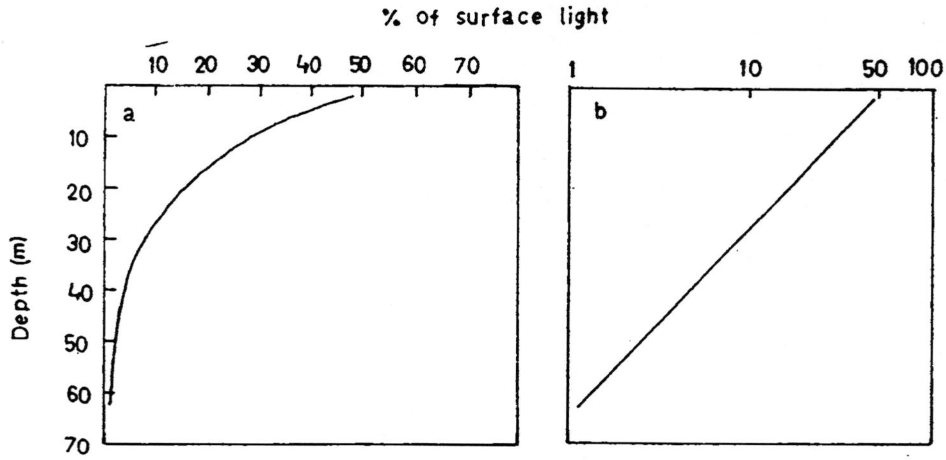

Fig. 5.3: The light extinction of lake Tahoe, California-Nevada on 27 August (1960) plotted, a, on a linear and, b, on a log scale (Goldman, 1967).

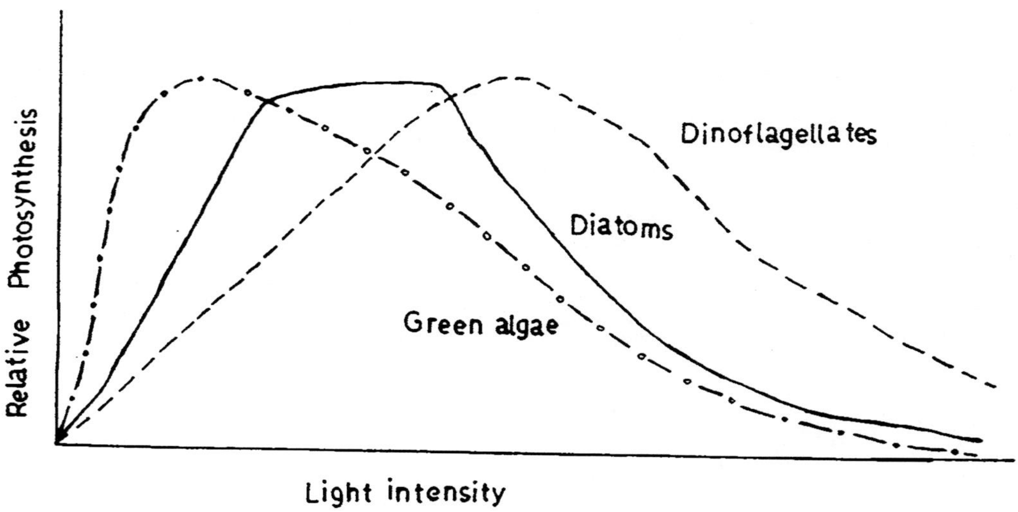

Fig. 5.4: Photosynthesis versus light intensity curves for three groups of phytoplankton.

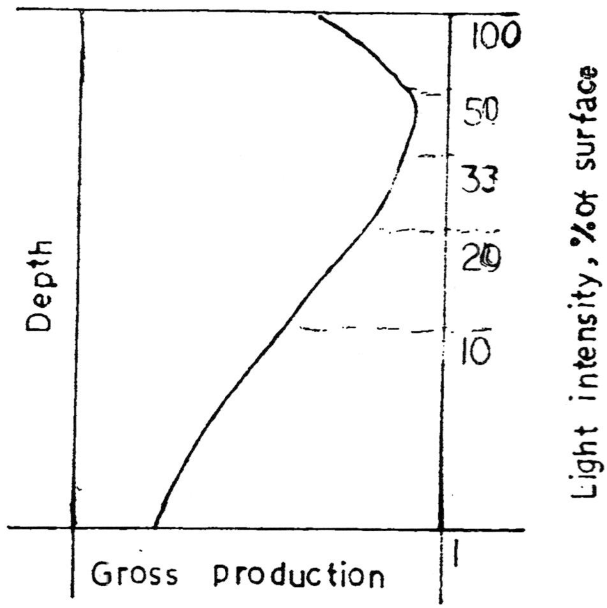

Fig. 5.5: Inhibition of photosynthesis in the surface layers of water as measured by gross production at different levels in the water column.

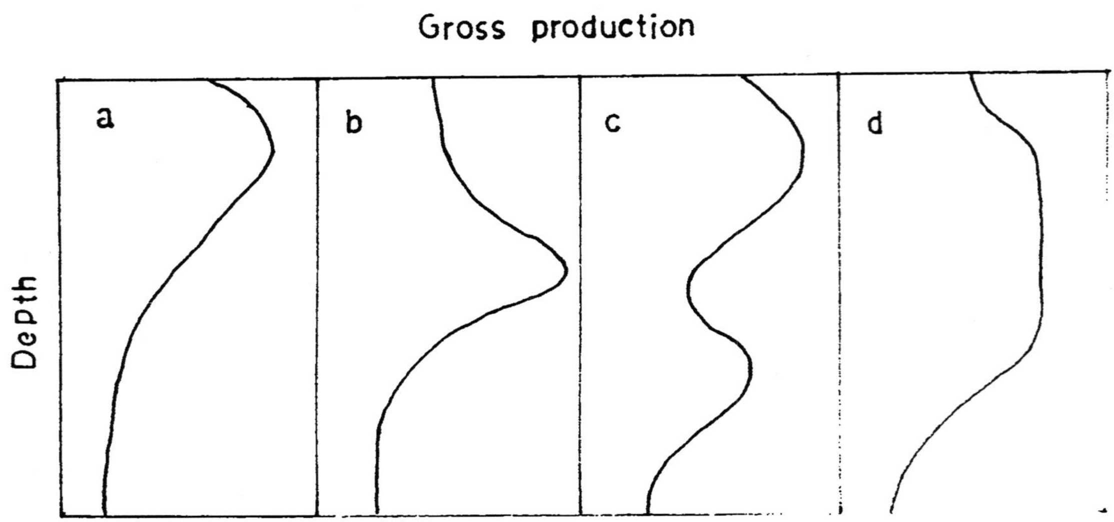

Fig. 5.6: Four types of vertical gross production profiles. a, in rich turbid lakes; b, in clean water lakes; c, in lakes where blue-green algae collect in the metalimnion; d, in some lakes where no distinct maximum occurs.

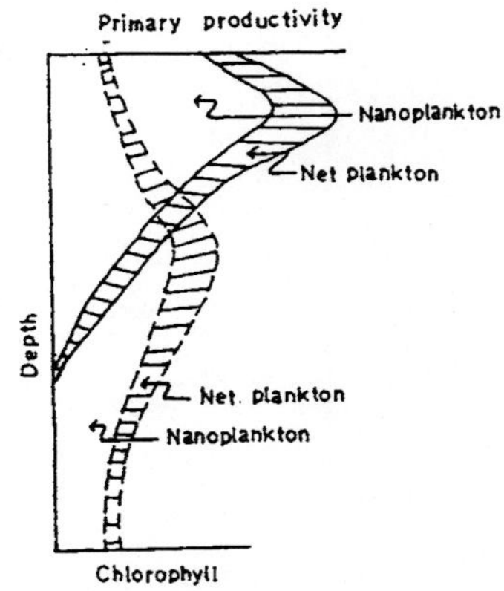

Fig. 5.7: The vertical distribution of primary productivity (continuous line) and chlorophyll a (dashed line) in the water column in lake. Each curve shows the larger contribution by the nanoplankton relative to the net plankton down the whole profile.

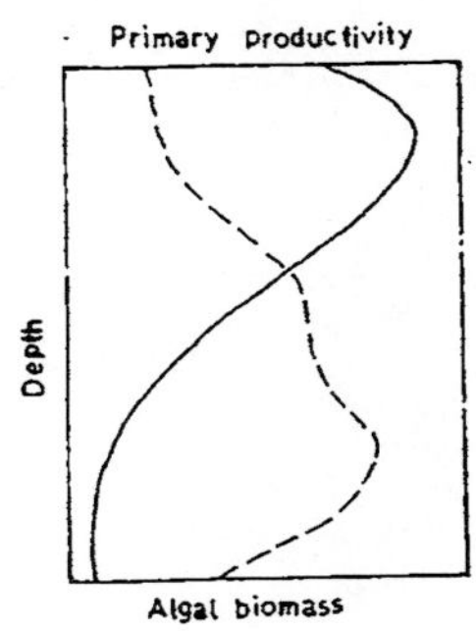

Fig. 5.8: Vertical profiles of total biomass (dashed line) and primary productivity (continuous line) in a lake.

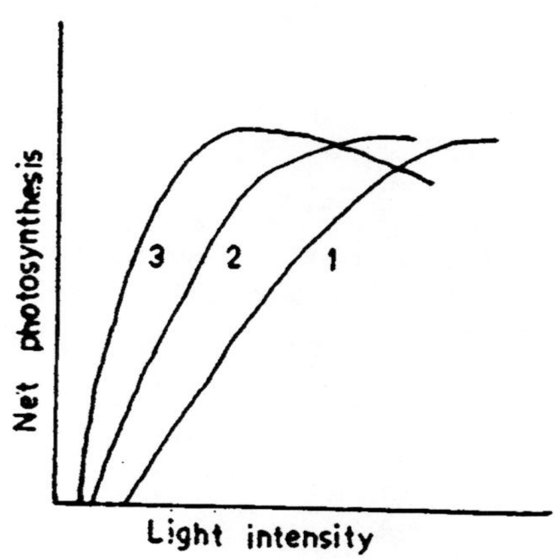

Fig. 5.9: Light intensity and relative rates of net photosynthesis in arctic phytoplankton from three different depths; 1, surface; 2, 25 m; 3, 50 m.

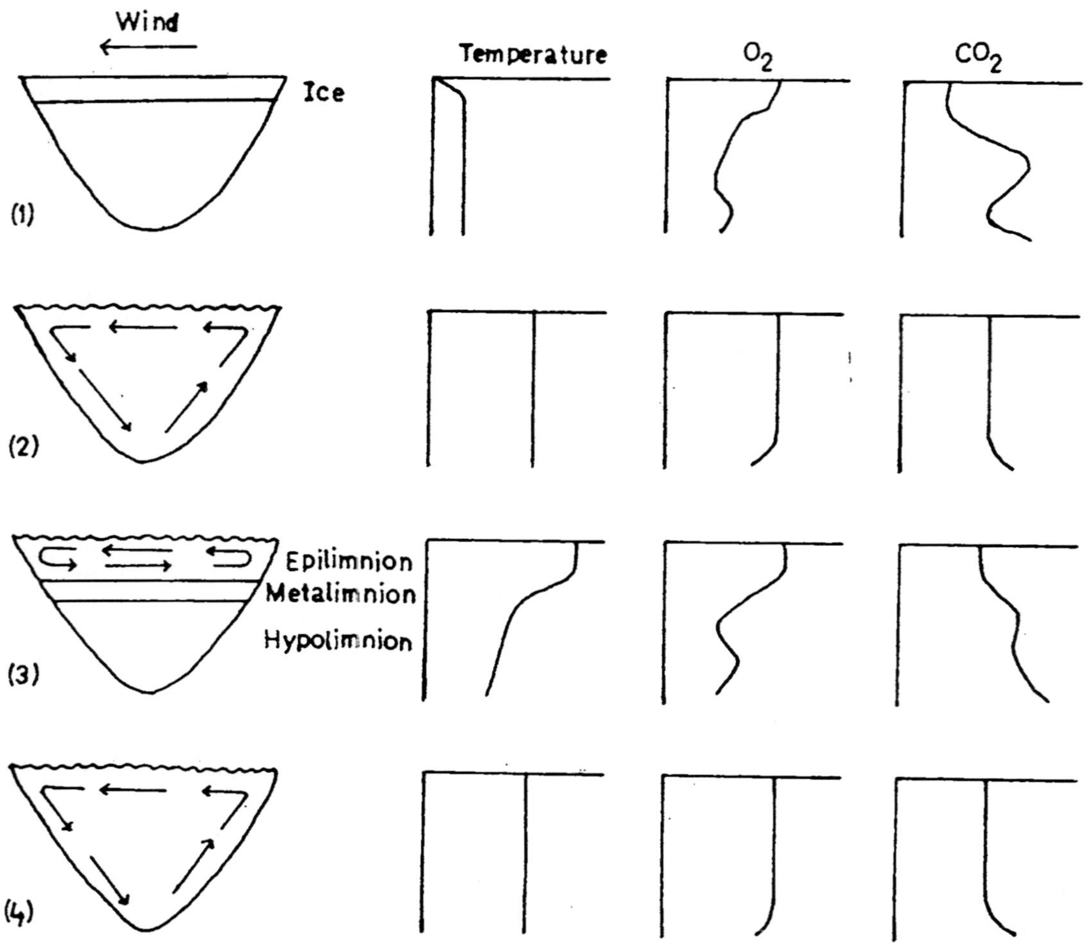

Fig. 5.10: Diagram showing seasonal circulation patterns in a temperate region lake and the variations of temperature, oxygen content and CO_2 content in water with depth at different seasons under different systems of stratification. 1, winter; 2, spring; 3, summer; 4, autumn.

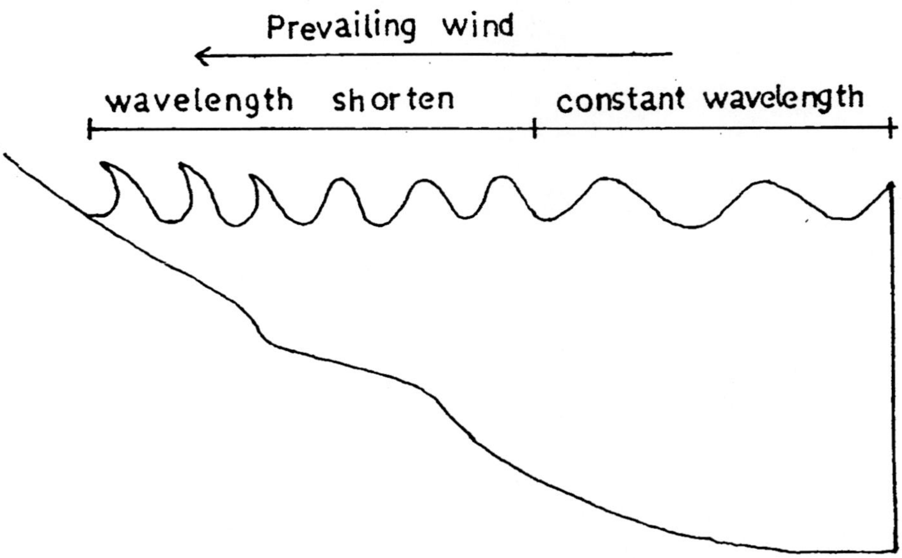

Fig. 5.11 Features of wave and change of wave form as it enters shallow water.

VI

CHEMICAL FACTORS AFFECTING THE FLORA

6.1 OXYGEN

Since oxygen does not become chemically bound with the water as does carbon dioxide, it can not be stored in this form nor can it be added to the water from dissolved rocks. The amount of oxygen in the water is only one-fortieth to one-twentieth of that present in an equal volume of air, when the two are at equilibrium and their partial pressures are the same.

6.1.1 Sources and distribution

The dissolution of oxygen from the atmosphere is the major source of oxygen in any water body. The rate at which the oxygen diffuses into the water from the atmosphere increases as the water turbulence increases. Although, the dissolution of oxygen from the atmosphere occurs at the surface, its downward movement into a water body requires mechanical mixing or mass flow of water. The rate of diffusion of dissolved oxygen through water is so slow that it is of no practical importance in transforming this element to the deeper strata of lakes.

The photosynthesis in submerged plants is another source of oxygen in underwater habitats. The peak of phytoplanktons photosynthesis usually occurs in subsurface waters, because full sunlight at the water surface is inhibitory to their photosynthesis (Fig. 5.5). In still waters with little mixing, this source may be of significant value in maintaining the animal's life at greater depth than it would otherwise be possible. Since, the presence of light is an all important factor in the photosynthetic production of oxygen, a daily, a spatial and a seasonal variation in the oxygen concentrations of aquatic environments may be expected. Enormous diurnal changes in oxygen concentrations in the water takes place due to algal photosynthesis and respiration. The rate of oxygen production by algae may be upto 20 times more in day time than the rate of oxygen uptake in respiration at night. The dissolved oxygen content is many times more near water surface and declines with increasing depth due to reduced photosynthetic release at decreasing light intensities.

The oxygen content of deep waters also depends upon the presence of absence of thermocline. Due to the presence of thermocline or metalimnion in stratified lakes, the surface source of oxygen is cut off and the oxygen content of hypolimnion is decreased (Fig. 5.10). The decomposition of organic materials at the bottom sediments also deplete the oxygen content of the hypolimnion. Therefore, the oxygen content of the hypolimnion is a reflection of the fertility of the lake and is inversely related with the amount of decaying organic matter.

The oxygen supply of deep waters is renewed during spring and autumn overturns (Fig. 5.10). If the lake is covered with the ice and snow during winter, a continuous reduction of dissolved oxygen content of the underlying water results, due to lack of exchange with atmosphere, low photosynthetic activity in dim light and continual respiratory removal.

The solubility of oxygen in the water is a function of its temperature. The lower is the temperature of a water, the greater will be the solubility of oxygen in that water (Table 5.1). The oxygen saturation capacity of freshwater at 0°C is 10.2 c.c/litre, but that at 25°C is only 6.0 c.c./litre. As the water warms up during summer, its capacity to hold oxygen is reduced and the dissolved oxygen is released into atmosphere. In warm water, the rate of respiration and decomposition of organic matter increases, which in turn decreases the dissolved oxygen content of the water and increases the risk of suffocation and fish-killing.

A very thick floating crust or scum of green or blue-green algae buoyed up on the water surface due to entrapped oxygen act as a barrier to the penetration of atmospheric oxygen as well as the light into the water and may suppress or decrease the photosynthetic release of oxygen of submerged plants and algae, and when the surface scum of algae die and settle to the lake bottom, its decomposition will further deplete the dissolved oxygen content of the water. The oxygen produced by the surface floating algae escape into the air.

A solid cover of free-floating plants such as *Lemna, Wolffia, Spirodela, Pistia* etc. on the water surface may also reduce the dissolved oxygen content of the water. The floating plants shade the water column and decreases the photosynthetic release of oxygen from submerged plants and also reduces the area of contact between water and atmosphere and thus reducing the dissolution of oxygen into the water. In addition, when their dead roots, leaves and other organic debris shed into water, they adds to the biological oxygen demand through their decay.

6.1.2 Influences

(i) Direct requirement by plants and animals

Because of their photosynthetic activities, higher plants and algae are practically independent of free oxygen in solution in daylight, but they require to draw upon the dissolved oxygen present in the water for their respiratory purpose in the night. Most of the dinoflagellates prefer well oxygenated water and are absent from ponds and lakes supplied with sewage. The lack of oxygen induce the shedding of flagella, immobility and encystment in algae. However, many diatoms are found to be present in highly organic polluted waters and are able to tolerate low levels of oxygen present in the water.

By increasing the dissolved oxygen content of the water, the submerged algae and aquatic plants can act as agents in determining the quantity and kinds of animals life present in any water body. The lowering of the oxygen content of the water by decay of dead organic matter in undisturbed lakes in hot weather can cause mass mortality of fishes. The oxygen also helps to prevent foul or septic conditions by favouring the activities of aerobic rather than anaerobic bacteria. In the presence of sufficient quantities of sewage, any dissolved oxygen present in the water is readily consumed and this generally limits the activities of aerobic bacteria. The products of aerobic decomposition process includes carbon dioxide, nitrate, phosphate, sulphate etc. while those of anaerobic decomposition process includes methane, sulphides, volatile acids etc. The former products are generally odourless, more soluble in water and are readily utilized by algae and other plants than the latter.

(ii) Availability of phosphorus to plants

In the presence of oxygen, iron is present in the oxidized state and forms relatively insoluble ferric phosphate with phosphorus, that precipitates out of solution, but under anaerobic conditions the iron is reduced and the corresponding phosphorus salt that forms i.e. ferrous phosphate is relatively more soluble. This means that in stratified lakes, the oxygen-rich epilimnion will have less available phosphorus than the oxygen-poor hypolimnion, and as summer progresses, the epilimnion will become poorer in phosphate as the ferric phosphate precipitates out and the dead bodies of plankton, plants and animals sink into hypolimnion carrying their phosphorus with them, and this coupled with the release from the sediments results into high concentrations of phosphorus in the hypolimnion by late summer. During winter, when lakes are ice covered in higher latitudes, the underlying waters will again become oxygen deficient, releasing more phosphorus and which will be available for algal growth during spring turnover.

(iii) Availability of trace metals to plants

The availability of trace metals to plants is also controlled by the oxygen distribution in the water. The iron and manganese are relatively insoluble in their oxidized states, but are more soluble in the anoxic and reducing environment of the bottom water and are recycled to the upper water during periods of mixing. During summer stratification, the iron and manganese are precipitated out in the oxygenated epilimnion and their concentrations are sharply reduced.

(iv) Nitrogen fixation

As oxygen interfer with the activity of nitrogenase enzyme-complex, it is thought that nitrogen fixation occurs in non-heterocystous species under anaerobic or microaerobic conditions. It has been suggested that the occurrence of trichomes of *Oscillatoria* in bundles may be one mean of achieving microaerobic environment. The central cells in the middle of such trichomes bundles are often discoloured and photosynthetically inactive and can perform nitrogen fixation.

6.2 CARBON DIOXIDE

6.2.1 Sources and distribution

Although the carbon dioxide is present in very low concentrations in the air, it is about 200 times more soluble in the water than is the oxygen, but the rate of diffusion of carbon dioxide in the water especially rich in minerals is 10,000 times slower than in the air. The water also contains large supplies of carbon dioxide from respirations, decay and underground sources. Unlike oxygen, carbon dioxide enters into a chemical combination with water to form bicarbonates and carbonates.

The carbon dioxide content of the water depends upon the temperature of water, depth of water, rate of respiration, decomposition of organic matter, chemical nature of the bottom and the geographical and physiological features of the terrain surrounding the water. The carbon dioxide content of the water increases with increase in temperatures and is generally higher in tropical waters than in temperate or polar waters. The vertical distribution of carbon dioxide in water column is closely linked inversely with oxygen levels. Both free and bound carbon dioxide are relatively low at the water surface and increases with water depth and are plentiful at or near the bottom (Fig. 5.10). A daily periodicity in the rate of carbon dioxide fixation by phytoplanktons has been noted with a maximum before mid-day and a minimum around 1900 Hrs.

6.2.2 Influences

Emergent or surface floating aquatic higher plants obtain carbon dioxide from the atmosphere, which enters the stomata on the leaves, while submerged plants must obtain the carbon dioxide dissolved in the water. Only those bodies of water abundantly supplied with carbon dioxide can support a luxuriant growth of algae. Although, an adequate supply of carbon dioxide is essential for development of a superabundant water bloom, an increase in carbon dioxide content, especially if rapid may seriously upset the physiology of fishes.

When carbon dioxide dissolves in the water, a certain percentage of its molecules react with water to produce soluble carbonic acid. Carbonic acid further dissociates into a hydrogen ion and an intermediately soluble bicarbonate ion, which may further dissociate into an another hydrogen ion and a nearly insoluble carbonate ion.

$$CO_2 + H_2O \rightleftharpoons H_2CO_3 \rightleftharpoons H^+ + HCO^-_3 \rightleftharpoons H^+ + CO_3^{2-}$$

$$\begin{array}{c} +H_2O \\ \updownarrow \\ H_2CO_3 + OH^- \end{array}$$

The carbon dioxide-carbonic acid-bicarbonate system functions as a buffer to keep the pH of water within a narrow range. It does this by absorbing H^+ ions in the water when they are in excess and producing more when they are in short supply. This is accomplished by shifting the above reactions to the right when there are too few H^+ ions, thereby producing more bicarbonate and carbonate ions, and to the left when there are too many H^+ ions, producing more undissociated carbonic acid and bicarbonate ion. Therefore, the availability of free carbon dioxide, bicarbonate and carbonate depends upon the pH of water. In acidic water, free carbon dioxide predominates, but it rapidly declines above neutrality, and at pH between 7 and 9 bicarbonate predominates, while above this level the carbonate becomes important (Table 6.1).

Although, all algae and higher plants are able to take up free carbon dioxide as the carbon source, those growing in hard or alkaline waters are able to assimilate bicarbonate as a carbon source, while those growing in soft or acidic waters are probably restricted to use free carbon dioxide as a carbon source. An utilization of bicarbonate ions, the overwhelmingly dominant form of dissolved carbon at high to neutral pH, as an alternative carbon source requires special adaptations by the plant cells. They must have either the ability to take up bicarbonate ions and convert them intracellularly into carbon dioxide by an enzyme carbonic anhydrase, or must promote extracellular conversion of bicarbonate ions into carbon dioxide near the cell surface either through an activity of extracellular pool of carbonic anhydrase or through transmembrane pH gradients. An algal species having such abilities has a great competitive advantage over those which do not have them, because free carbon dioxide is frequently a limiting factor for plants growth in dense strands of aquatic vegetation or algal bloom.

When bicarbonate is converted into carbonate, the pH of water rises and the carbon dioxide released is utilized for algal photosynthesis. The degree of pH change per unit of CO_2 removal depends upon the buffering capacity of the water. Thus one unit of CO_2 removal by photosynthesis will bring about a greater pH increase in soft water lakes than in well buffered hard water lakes or in sea water.

$$Ca(HCO_3)_2 \rightleftharpoons CaCO_3 + H_2O + CO_2$$

The carbonate formed after bicarbonate decomposition frequently precipitate out as calcium carbonate and is deposited on the surface of submerged plant parts, giving them a coarse limey texture. Such calcium carbonate deposition has been found to occur on algae like *Chara, Chaetophora, Oocardium, Rivularia, Nitzschia* etc. and can increase their survival against grazing animals. However, the absence of calcium carbonate deposition on most of the algae may be due to release of some extracellular substances like polyphenol which may inhibit such deposition.

The calcium carbonate may also accumulate, as to make conspicuous marl deposit, on the bottom of some lakes, and when marl formation becomes considerable, the fertility of lake decreases with a consequent decrease in animals life including those inhabiting the bottom. But when there is sufficient free carbon dioxide present in the water derived from other sources, the calcium carbonate (lime stone etc.) is converted back into bicarbonate and marl does not form.

At high pH, carbonates are prevalent, one often finds precipitation of phosphorus in association with carbonates and as a result it is difficult to tell that whether a change in species composition is due to low phosphorus or a shift in the form of carbon. It is to be noted that when both the bound carbon dioxide (carbonates and bicarbonates) and pH of water are high, the blue-green algae predominates in the water, but with lowering the pH and decrease in bound carbon dioxide, there is an increasing predominance of green algae.

6.3 NUTRIENTS

6.3.1 Occurrence

In lake water, the nutrients levels are largely determined by the nutrients status of the surrounding land or the river draining into it. The percolation of rain water through soils yields solutions of increasing concentrations and their exact composition varies with the geology, soil type, vegetation, evaporation etc. During this solution process, four cations (calcium, magnesium, sodium and potassium) and three anions (bicarbonate, sulphate and chloride) are greatly increased, so that they make up the bulk of ions in freshwater bodies (Table 6.2). In addition, there is usually a variable amount of silicic acid present.

The falling rain may contain as much as 30 to 40 ppm (mg/L) of nutrients and its runoff over uplands may dissolve more nutrients into it before it drains into a stream or a lake. The rain water which run off over siliceous or sandy soils may contain about 50 to 80 ppm of dissolved minerals, while those which run off over more fertile calcareous soils may contain upto about 350 ppm of dissolved minerals. In some saline lakes, the total dissolved salts may exceed 100,000 ppm. Although, less precise and exhibiting greater ranges, the ratio of major elements in different freshwater bodies varies only slightly, but with a tendency for alkali content to rise in saline lakes, and those of calcium and magnesium in hard water lakes (Table 6.2). The phosphorus and nitrogen are most likely to be deficient in almost all lakes (particularly in those having no sewage input). The nutrients tend to accumulate in deep waters, but they are brought back to surface during spring and autumn overturns.

6.3.2 Chemical stratification of water

The nutrients are present in dilute solutions at spatially and temporally varying concentrations in lakes. The lake water is chemically stratified vertically during summer stratification when the thermocline cuts off the supply of nutrients from hypolimnion into epilimnion. The accumulation, decomposition and mineralization of dead organic matters at the bottom, the entry of basal springs and the deoxygenation and build up of hydrogen sulphide at the bottom sediments, all are responsible for chemical stratification of water. The very little ammonium is present in the surface waters unless they have a sewage input, but ammonium accumulation does occur in poorly aerated deep waters as a result of decomposition of proteinaceous organic matter as well as nitrate reduction (Table 6.3).

In surface waters, the low concentrations of inorganic nutrients does not necessarily indicate their total scarcity, since they are in such demand that they are rapidly taken up by primary producers and will be out of circulation for some times unless released after the death and decomposition of primary, secondary and tertiary producers.

The seasonal fluctuations of the concentrations of the nutrients like magnesium, calcium, potassium, sodium, chloride etc present in the water are slight (and their concentrations can therefore be used to characterize a freshwater body), but those of others like nitrate, phosphate, silicate, iron, copper, zinc etc. vary very much because of their heavy demand by primary producers.

6.3.3 Nutrients cycle

The nutrients are as important to algae as is the solar energy, but differ from the latter, that they are not used up but are cycled through living organisms and environment and are not totally lost in the biosphere. Their greatest quantity is locked up for a long time in rocks and also in lake sediments. The nutrients limitation is likely to occur only in the surface waters and not in deep waters.

In nutrients cycle, there is major reservoir or pool of elements from which they are continually moving in and out as they passes through organisms. Algae incorporate large quantities of inorganic nutrients even from great dilutions in the water (either from active absorption through cell membranes, adsorption on to exposed surface area or from engulfment). These nutrients are to a large extent deposited on the sediments after the death of algae or after passing through animals. Animals in general, excrete and secrete during their life span many times the amount of soluble nutrients as would be present for microbial decomposition after their death. The excreted and secreted dissolved inorganic and organic compounds of nitrogen and phosphorus (and of course CO_2) can be used up directly by primary producers without any further chemical breakdown by bacteria. Some benthic animals are detritus feeders and obtain most of their food energy from eating bacteria, protozoa and other organisms associated with ingested detritus. The dead organic matters of plants, animals and microbes are broken down by autolysis and are decomposed enzymatically by bacteria and related microoganisms and the inorganic nutrients are released into the abiotic pool (Fig. 6.1). The maintenance of life depends upon the continual recycling of inorganic materials involving the decomposition of organic matter, release of inorganic nutrients and their absorption by primary producers again. The decay of organic matters involves the simultaneous mineralization of carbon, nitrogen, phosphorus, sulphur etc. and thus all of them are interlocked and interdependent in nature. The major part of organic matter is decomposed rapidly within few days while still in the upper layers or in shallow sediments, but a small part of organic matter accumulated on the bottom sediment does not decompose completely (because of low oxygen availability or their less degradability) and may exist there either in the form of humus, fossil, skeleton or fuel, and the inorganic nutrients incorporated in them are released very slowly in the abiotic pool either through volcanic activity, slow erosion, slow decomposition or burning.

6.3.4 Factors affecting nutrients uptake by flora

Although, oxygen and carbon dioxide are taken up passively by algae and other aquatic higher plants, except at high pH when active transport is necessary for the uptake of bicarbonate ions at rates sufficient to support net photosynthesis : other inorganic nutrients are generally taken up in ionic forrn by way of specific energy mediated, membrane bound permease systems which functions to provide elevated intracellular concentrations of these ions as substrates for various assimilatory enzymatic processes. The following factors may influence nutrients uptake in algae.

(i) Light

Nutrients uptake is often, but not always, stimulated by light, sometimes as much as 10-15 folds over that of dark uptake rate. Light can influence nutrients uptake via photosynthesis which provides energy necessary for active uptake of ions, or carbon skeletions which act as sink for incorporated ions, or stimulates protein synthesis and growth which results into increased nutrients uptake. Therefore, uptake of nutrients by algae exhibit diurnal periodicity.

(ii) Temperature

It will influence the rate of diffusion and carrier mediated uptake of nutrients by a factor of about 2 for each 10°C rise upto some optimum temperatures and thereafter the rate of uptake declines.

(iii) Extracellular nutrients concentrations

The rate of nutrients uptake is related with their extracellular concentrations. This implies that at low extracellular nutrients concentrations, the rate of nutrients uptake increases rapidly with an increase in external concentrations, but at higher extracellular concentrations, an increase in nutrients concentrations adds progressively less to their uptake rates until a high point is reached at which their uptake is virtually constant and will become be independent of their external concentrations.

(iv) Competing ions

The uptake of ions from the external medium is also influenced by the concentrations of competing ions which may engage or suppress the common carrier of enzyme system.

(v) Water turbulence

In low turbulence regions, the algae may exhibit indications of nutrients deficiency even when the external nutrients concentration is moderate, but they will grow profusely in turbulent but low nutrients containing medium. The water currents flowing past the algae will renew the boundary layer by removing the algal extracellular substances and providing more nutrients to algae.

6.3.5 Adaptations in algae in response to nutrients availability

Some algae are able to scavenge nutrients from their extremely low external concentrations and are able to store them in excess of their metabolic need and to use them for their growth and reproduction when needed and thus have a competitive advantage over others.

(i) Luxury consumption

Many algae are able to take up large amounts of various nutrients beyond their immediate metabolic need under very favourable nutrients-enrich conditions and can store them either in the cytoplasm, cell vacuoles or in the cell wall and can utilize them when external nutrients concentrations becomes growth limiting and thus are not dependent for some times upon the suboptimal external nutrients concentrations.

An important difference between algae and vascular plants is that the former can store large amounts of polyphosphate bodies (formed after enzymatic polymerization of assimilated orthophosphate) but not the latter. It is believed that polyphosphate bodies functions as deposits of phosphorus and can be easily used to continue growth for some times under phosphorus depletion.

Many blue-green algae also appear unique in their ability to store nitrogen as structured granules of cyanophycin (a polymer of amino acids arginine and aspartic acid containing high N to C ratio) and also nitrogen containing pigment phycocyanin and by implication phycoerythrin.

Some algae have also been reported to synthesize and store large quantities of organic carbon compounds in presence of excess solar radiation and can utilize these reserves when needed at the time of growth. This important ecological strategy may enable them to persist and even flourish during periods extremely stressful to their survival such as during winter stratification in polar lakes.

(ii) Enzyme secretions

When there is a nutrients shortage in the water, many algae are able to secrete enzymes, which are able to structurally modify or break the organic compounds and can release the limiting nutrients present in them. For example, during phosphorus deficiency, some algae secretes an enzyme alkaline phosphatase, which appears to be either located near the cell surface or in the surrounding medium and speed up the recycling process by breaking down the organo-phosphorus compounds enzymatically and releasing the phosphate without their complete bacterial decomposition and mineralization and thus lay a real advantage to the algae. A positive relationship with light has also been observed to occur for the activity of this enzyme and thus the activity of this enzyme being higher in the upper euphotic zone than in the lower euphotic zone.

The production of extracellular enzyme is not only restricted to phosphorus procurement but also appears to occur for other elements like nitrogen. Under nitrogen deficiency some algae produce extracellular deaminases which allow them to utilize amino groups of some amino acids.

(iii) Increase in surface area

A positive relationship exist between the surface area of algae and the nutrients uptake. Small algae with high surface area to volume ratio can persist and grow in nutrients poor oligotrophic waters. There is an increasing preponderance of nanoplanktons along a decreasing gradients of nutrients concentrations. In some algae, hyaline hairs are formed in order to increase nutrients uptake under conditions of low nutrients concentrations and high light intensities. If an alga growing in nutrients poor medium and bearing hairs is transferred to nutrients rich medium, it will shed the hairs.

(iv) Cell movement

Forced convection produced by cell movement ensure the continuous removal of nutrients-poor cell contact layer formed around the static cell with the fresh nutrients-enriched waters and thus increase the opportunities of nutrients uptake. A stationary cell will always be at a disadvantage because of the formation of a nutrients-poor cell contact layer which slow down the nutrients uptake and also build up a shell of extracellular products having pH different from the surrounding medium.

Dinoflagellates and other flagellates like *Mallomonas, Chroomonas* etc. undergo diurnal vertical migrations which helps them to exploit nutrients enriched deep water. They migrate towards nutrients-enriched waters of hypolimnion at the night and return to the upper enlighted water at morning and thus have obvious ecological advantage over others with respect to seeking light at day time and obtaining most of the nutrients at night.

6.3.6 Nutrients required by aquatic flora

Algae requires a variety of inorganic and organic nutrients for their growth and reproduction. Some of the essential inorganic macro-and micronutrients required by different algae are hydrogen, oxygen, carbon, nitrogen, phosphorus, calcium, magnesium, potassium, sodium, sulphur, silicon, iron, manganese, boron, zinc, copper, chloride, molybednum, cobalt etc. These nutrients are derived either from water, dissolved gases, rock weathering, soil leaching, fertilizers input, animals excretion and sewage discharge. They are taken up by algae and aquatic higher plants without any further chemical breakdown by the bacteria.

Some algae also require for their growth, the presence of organic compounds like carbohydrates, amino acids, fatty acids, organic acids, vitamins and hormones, present in very low concentrations in the water as secretory or excretory products of actively growing bacteria, fungi, plants and animals or as decomposition products of dead organic matters. Unlike bacteria which have an uptake mechanism adapted for low concentrations of these organic substances, algae do not have an enzyme permease and other extracellular enzymes to the same degree as they are present in the bacteria and requires excessive energy to concentrate these substances present in very low levels in the water.

Any or all of these substances may act by their presence or absence or quantity in a critical way to determine the density and diversity of algal species. Since algae are physiologically quite heterogenous as compared to vascular plants, it is impossible to make any generalization about nutrients requirement of algae. Some algae demands certain elements much in excess than what others requires. An element essential for one algal species may not be for others. For example, diatoms requires silicon for their cell wall formation and growth, similarly many blue-green algae requires more nitrogen and phosphorus for their growth then others. The ammonium nitrogen is the preferred source of nitrogen for euglenoids. The diatom *Melosira distans* prefers waters of low nutrients concentrations (soft waters), while *M. granulata* prefers those of high nutrients concentrations (hard waters).

(i) Inorganic macronutrients

Hydrogen

Using energy from the sun, the water molecules get split into oxygen and hydrogen in the plants. The hydrogen is entrapped and used in the photosynthesis, but the oxygen is released. The hydrogen is never a limiting factor for the growth of algae and aquatic plants.

Oxygen

(see section 6.1)

Carbon

The carbon exist in the four different forms in the water.

(a) *Dissolved inorganic carbon* It exist in the form of dissolved carbon dioxide and bicarbonate ions and in some waters it may be the largest pool present.

(b) *Particulate inorganic carbon* It is the smallest pool existing in the form of calcium carbonate precipitated out by algae, animals or by physical means.

(c) *Dissolved organic carbon* A very large pool of dissolved organic substances containing carbon molecules secreted or excreted by bacteria, fungi, plants or animals is present in the water. Perhaps one-fourth of the carbon fixed by photosynthesis is released into water as dissolved organic matter leaked from the cells. Free living bacterioplanktons are mostly heterotrophic on dissolved organic carbon compounds and convert a substantial amount of them into particulate organic carbon of their cells. Thus the carbon leaked from phytoplanktons is not really lost but is recycled back into food chains. The bacteria in turn are consumed by different nanozooplankton and thus can channel a significant amount of energy into particulate food webs.

(d) *Particulate organic carbon* It is present in the form of particulate living and dead materials of all plants, animals and bacteria present in the water.

The carbon reservoir in the water exist in the form of carbonic acid-bicarbonate-carbonate system. It is fixed into organic carbon compounds by algae and aquatic higher plants through photosynthesis, transferred to animals through herbivory and predation and returned to the reservoir via respiration and bacterial decomposition. A nonrenewable loss occurs when heavy carbonate materials are deposited in deep water as calcareous sediments (Fig. 6.2).

Nitrogen

The elemental nitrogen is present abundantly both in the air as well as in the water, but there is no evidence that algae other than few Cyanophyceae members are able to utilize it directly. The inorganic and organic nitrogenous compounds present in the freshwaters varies greatly and are derived either from inflow through agricultural lands containing nitrogenous fertilizers, river input, sewage discharge, animal's excretions or decomposition of nitrogenous organic compounds. The algae can utilize inorganic nitrogen compounds such as nitrate, nitrite and ammonium as well as organic nitrogenous compounds like urea, uric acid and amino acids for their nitrogen need and this perhaps explain that why nitrogen is only rarely limiting to algal production in waters.

Although, uptake of ammonium generally exceeds to that of nitrate at all nitrogen levels normally present in the water, at high concentrations it is toxic to most of algae. The algae grow optimally at high pH if nitrate rather than ammonia is the nitrogen source, because ammonia often forms ammonium hydroxide at high pH, which is toxic to most algae. Urea proved to be a comparatively inferior nitrogen source for algae.

In thermally stratified lakes, the nitrogen is primarily present as ammonium in the lower waters, as nitrate in the upper waters and as nitrite (usually released by phytoplanktons at low light levels) near the base of euphotic zone. In the water, ammonia and urea originates mostly from the decomposition of nitrogenous organic compounds and animals excretions.

The four main process of nitrogen cycle (Fig. 6.3) largely controlled by aeration are as follows:

(a) *Nitrogen fixation* It is the process of reduction of dinitrogen gas to ammonium by some nitrogen fixing bacteria like *Azotobacter* and *Clostridium* and blue-green algae like *Anabaena. Aphanizomenon* etc. This process requires anaerobic condition or at least reduced oxygen tension.

(b) *Nitrification* The ammonium is oxidized to nitrite by bacterium *Nitrosomonas* and then to nitrate by *Nitrobacter*. In aerobic freshwater, it is a well established phenomenon.

(c) *Nitrate reduction* In deep waters, where decomposition of organic matter predominates, nitrate is reduced to nitrite by bacteria *Thiobacillus* and *Pseudomonas* and then to ammonia by Desulfovibrio.

(d) *Denitrification* Nitrite is reduced to gaseous nitrogen by facultative anaerobes *Pseudomonas.*

Since inorganic salts of nitrogen are very soluble in the water and are not readily bound to the exchange complexes of clay minerals and organic matter in the sediments, a greater percentage of them is found in the dissolved form in the water and not as an exchangeable form in the bottom mud (while opposite is the case with phosphorus). Since nitrogen is generally more soluble and more mobile, it leach out more readily and washes into and out of the water courses with greater ease.

The nitrogen is a key component of all amino acids and hence of proteins and nitrogenous bases (purines, pyramidines) of nucleotides, porphyrin ring of chlorophyll and cytochrome and of some toxins. Some phospholipids also contain nitrogen. Phytoplanktons growing in nutrients unlimited environments typically have a C : N : P atomic ratio of 106 : 16: 1, the so called *Red field ratio.* Thus next to oxygen, hydrogen and carbon, an element present in the greatest amount in the algal cells without mineralized wall is the nitrogen. The large departure from the *Red field ratio* suggest nutrients limitation.

Phosphorus

In an unpolluted water, the primary source of phosphorus is the weathering of phosphate rock from which it leach out very slowly. The metabolic acids released by microorganisms can also dissolve the inorganic phosphorus present in the rock. The phosphorus almost entirely present as orthophosphate ions (HPO^{-}_{4}) exist in the lake either dissolved in the water, adsorbed on the suspended or sedimented particulate organic matter, contained in living and dead organic matters or as precipitated inorganic salts on the sediments. The phosphorus concentrations are much higher in the sediments than in the water.

The phosphorus is an essential constitutent of sugar phosphates, nucleotides, phospholipids and adenosine phosphates (AMP, ADP, ATP, NADP etc, which are crucial for energy metabolism). Phosphate seems to be limiting to some extent in nearly all freshwaters and its addition can increase the growth of algae (particularly blue-green algae like *Oscillatoria, Phormidium, Lyngbya* etc. and green algae like *Chroococcus, Cladophora, Chaetophora, Stigeoclonium* etc.).

The oligotrophic lakes frequently have less than 0.001 mg/L of phosphorus and the algae are capable of taking up available phosphorus very quickly. Since internal phosphorus levels are always likely to be higher than external levels, the phosphorus requires active transport. As algae removes the soluble form of phosphorus from the water, more phosphorus is releashed from the bottom sediments to restore the equilibrium. Lowering the pH of the water results into release of more phosphorus from the sediments into water. Since ferrous phosphate is considerably more soluble than ferric phosphate, more phosphorus will be found in solution in oxygen-poor hypolimnion than in oxygen-rich epilimnion. The dead bodies of plants and animals sink into hypolimnion and carry their phosphorus with them. This coupled with the release from the sediments results into high concentrations of phosphorus in the hypolimnion. When water overturns in the spring and autumn, the phosphorus accumulated in the hypolimnion will be available to plants growing in the upper water.

The decay of nucleic acids and phospholipids is quite rapid. The phosphorus present as phosphate in the organic matter is merely released by the breakdown or autolysis of carbon skeletion within few days under sterile conditions (but nitrogen compounds remains there for more time). Much of phosphotrus is present in the bones of animals in the form of tricalcium phosphate and the phosphorus loss occurs when various skeletal materials sink into deep sediments. The bones of dead animals provide an excellent surface for the attachment of bacteria. The bacteria produce enough acids to have solvent action on the phosphate present in the bone. The parts of phosphate present in the bones of animals may also be dissolved by acidic digestive juices of carnivorous animals (Fig. 6.4).

Calcium and Magnesium

The calcium and magnesium contents in the water varies widely depending upon their contents in the soils. The waters in the limestone regions will be very hard with high levels of calcium and magnesium, while those lying in watersheds having acidic soils will be soft with low levels of these two elements.

The calcium is required in very small quantities and is used for the maintenance of cellular membranes, enzyme activation and ion transport, while magnesium is a constituent of chlorophyll and is involved in the aggregation of ribosomes into functional units, enzyme activition, ion transport and formation of catalase enzyme in the plants. Diatoms predominates in the calcium-rich waters.

The calcium bring about the precipitation of many heavy metals and phosphate and also is known to antagonize the toxic effects of various metals probably by inducing their precipitation. A pH change towards alkaline side due to photosynthesis in the epilimnion may precipitate soluble calcium bicarbonate into calcium carbonate and which may settle to lake bottom. At high pH, when carbonates are prevalent, one often finds precipitation of phosphorus along with such carbonates. The calcium is very important in the carbonate-bicarbonate buffering system. In calcium-poor waters, one typically finds lower pH and often fairly large amounts of iron and other trace metals.

Potassium

It is rarely limiting to the growth of algae and aquatic higher plants and is required in greater than trace amounts. It can not be replaced either by sodium or other elements and is therefore an essential element. Its functions includes osmotic regulation, pH control, protein conformation and maintenance of the electrochemical environment of the cells.

Sodium

It is not generally as an absolute requirement for the majority of algae, but it is required to maintain an exponential growth of some blue-green algae like *Anabaena variabilis, Anacystis nidulans, Nostoc muscorum* etc. An increase in blue-green algal growth in eutrophic waters may in part be associated with an increase in sodium content.

Sulphur

It is not a limiting nutrient for many algae under natural conditions. The main sources of sulphur in the freshwaters are the rain water and leaching of the soil. The most abundant form of sulphur in the water is sulphate, but in reducing conditions (e.g. in the hypolimnion of some lakes) it may be present as hydrogen sulphide. The hydrogen sulphide is produced by anaerobic decay of organic matter and from sulphate reduction by *Desulfovibrio.*

The photosynthetic sulphur bacteria *Chlorobium* and *Chromatium*, occurring mostly in the shallow waters provided with sufficient light and having anaerobic conditions, oxidizes the hydrogen sulphide in order to reduce carbon dioxide and produce sulphur. (see p. 63).

Thiobacillus spp present in the water can oxidize the elemental sulphur, sulphide, thiosulphate or tetrathionate to sulphate. The sulphates are utilized in the algal metabilism, but sulphides are often toxic.

The sulphur is an essential constituent of proteins containing amino acids methionine, cysteine and cystine and some coenzymes. Some dinoflagellates use organic sulphur such as cysteine and methionine for their growth. There is an apparent involvement of sulphur compounds in the uptake of silicon (an element essential for diatoms cell wall).

Silicon

The weathering of rock fragments is the major source of soluble silicon in the water. The solution of siliceous minerals enters into a water body either through surface runoff or percolation of water into the soil. The pre-existing diatoms frustules containing silicon undergo dissolution or mechanical breakdown during their long descent in the water column, the grinding movement of the sediments or during feeding and excretory processes of the grazing animals. The silica walls of dead diatoms are readily dissolved in very alkaline solutions and thus silicate is much more available in alkaline water than in acidic water.

The most important form of soluble silicon present in the water is undissociated orthosilicate ($Si(OH)_4$). In lake waters, the dissolved silicon content varies from undetectable to 400 µmol/L. The surface waters always contains less silicon than deep waters. The silicon is an essential and often a limiting nutrient for diatoms cell wall and chrysophytes scale formation. The silicon uptake is an energy requiring process and its prolong starvation lead to an accumulation of diatoms daughter cells remains united as no cell wall is formed. The rapid growth of diatoms and slow recycling of silicon may lower the Si: P ratio in the water in favour of other taxa, unless the deep mixing of water bring more silicon from lower water to upper water.

(ii) Inorganic micronutrients

The micronutrients or trace elements essential for the growth of algae are iron, manganese, boron, zinc, copper, chlorine, molybdenum, cobalt, vanadium etc. They are required in extremely small amounts and their slight excess can be toxic. Most of them are more soluble in acidic waters and at low redox potentials.

Iron and Manganese

Their distribution in the water are strongly controlled by the oxygen distributions. These metals are relatively insoluble in their oxidized states but are much more soluble in anoxic reducing environment of the bottom sediments and are brought to the surface waters during water overturn. During summer stratification, they are precipitated out in oxygenated epilimnion and their concentrations are sharply reduced. The concentrations of both iron and phosphorus are closely associated because both of them are more soluble under acidic conditions.

All algae have an unquestioned requirement for iron. It act both as a structural component and also as a cofactor for enzymatic reactions. The oxidation-reduction reactions including those of electron transport chain in both photosynthesis and respiration are most commonly associated with iron containing systems. The ferredoxin and some cytochromes are well known examples. The deficiency of iron lead to bleaching of cells since it is involved in the chlorophyll synthesis. Most algae grew best when iron content of water is between 0.2 to 2 mg/L. The iron is also used by many algae (e.g. desmids) in their wall synthesis. The manganese is required for photosystem II reactions and in manganese deficient cells, the photosynthesis is lowered.

Boron, Zinc, Copper and Chlorine

It is presumed that they are present in sufficient quantities in most freshwaters and are rarely limiting to algal growth particularly in acidic waters but may be limiting at high pH situations.

The deficiency of boron results into loss of pigments and reduction of growth in *Nostoc muscorum*. The diatoms usually have high activities of enzyme lactate dehydrogenase with zinc as an essential constituent. The copper is utilized in the electron transport system and is a constituent of plastocyanin and a cofactor for several enzymes. The chloride appears to be an essential factor for the photophosphorylation reactions.

Molybdenum, Cobalt and Vanadium

The molybdenum along with iron is a constituent of nitrogenase enzyme complex which is required for nitrogen fixation in blue-green algae. The cobalt is often present in the water as cyanocobalamin and constitute the 4 percent of the weight of vitamin B_{12} and thus appears to be an essential algal micronutrient. The vanadium seems to be associated with the process of photosynthesis and is required in very small amounts of about 20 μg/L to stimulate algal growth.

6.4 HYDROGEN ION CONCENTRATION

The acidity or alkalinity of any water depends upon the ratio of H^+ (or hydronium, H_3o^+) and OH^- ions. When the number of H^+ ions is equal to that of OH^- ions, the pH value will be 7 and which represent absolute neutrality. Most unpolluted lakes and streams have pH values ranging between 6.0 to 9.0, but extreme pH values of 1.7 and 12.0 may occasionally occur. In some water bodies, the pH values may fluctuate considerably. The hydrogen ion concentrations increases with active decomposition of organic matter. Dystrophic lakes, volcanic lakes and bog lakes, all are acidic in nature, whilst desert lakes are usually alkaline in nature.

6.4.1 Diurnal variations in pH

Aquatic algae and plants tend to lower the pH of water by releasing carbon dioxide during night, but will raise the pH of water by abstractions of carbon dioxide from bicarbonates for photosynthesis during day time. The actual fluctuations in pH values depends upon the density of algae and plants and also upon the buffering capacity of the water. In poorly buffered waters (generally those with low levels of carbonates, bicarbonates and phosphates), aquatic plants can cause a pH variation of 3 full units between sunrise and sunset. A soft-water pond may have a pH of 6.5 at sunrise and a pH of 9.5 at sunset and such extreme pH variations in such a short period of time may easily affect the organisms and perhaps even limit the species composition.

6.4.2 pH optima of different algae

Majority of algae grow best in water at or near neutral pH, but a considerable number of blue-green algae favour high pH values. The optimum pH for the growth of *Microcystis* and *Gloeotrichia* have been found to be about 10.0. Most of the freshwater dinoflagellates are also found to be alkalinophile. However, low pH oligotrophic lakes are usually dominated by desmids, cryptophyceae, and diatoms.

6.4.3 Availability of gases and nutrients to algae

The measurement of water pH may serve as an index for other environmental factors such as the amount of dissolved carbon dioxide (with which it varies inversely), dissolved oxygen (with which it varies directly) and nutrients. At low pH values, the molecular carbon dioxide is the principal form in which the inorganic carbon enters into algal cells, while at high pH values the algal growth is dependent on bicarbonate and carbonate and not on carbon dioxide (Table 6.1).

The release of phosphorus from the sediments is also dependent upon many factors like pH and oxygen levels in the water. The lowering of pH results into release of more phosphorus from the sediments into water, but raising the pH may tie up phosphorus to the sediments. At high pH, when carbonates are prevalent, one often finds precipitation of phosphorus in association with such carbonates.

Similarly, most of the trace elements tend to be more soluble and becomes more available to algae at low pH and low redox potentials. Acid waters and waters rich in humates usually contain more iron than alkaline waters. Zinc, copper and manganese becomes growth limiting in alkaline waters but never in acidic waters. However, silicate is much more available in alkaline waters than in acidic waters.

At pH above 7, the ammonia often forms ammonium hydroxide, which is toxic to ost algae. Thus, at high pH, nitrate rather than ammonia is the better source of nitrogen, at at low pH, either of them can be utilized for the growth of algae.

Table 6.1 Proportions of carbon dioxide, bicarbonate and carbonate ions in water at various pH values (Hutchinson, 1957).

pH	*Total free carbon dioxide*	*Bicarbonate ions*	*Carbonate ions*
4	0.996	0.004	1.25×10^{-9}
5	0.962	0.038	1.20×10^{-7}
6	0.725	0.275	0.91×10^{-5}
7	0.208	0.792	2.6×10^{-4}
8	0.025	0.972	3.2×10^{-3}
9	0.003	0.966	0.031
10	0.000(2)	0.757	0.243

Table 6.2 Chemical composition of various waters. Data from Gorham (1955), Hutchinson (1957), Livingston (1963), and Visser and Villeneuve (1975). Dashes indicate that values were not given.

	Rainwater (ppm)	*Riverwater (ppm)*	*Wisconsin lakes (minimum concentration (ppm)*	*Larger East African Rift lakes (ppm)*	*East African saline lakes (ppm)*
Chloride	0.2-12.6[a]	7.8	0.1	29.9	16385
Sodium	0.2-7.5[a]	6.3	0.13	105	18261
Sulphate	1.1-9.6[b]	11.2	0.75	39.5	2175
Magnesium	Nil-0.8[a]	4.1	0.5	33.7	0.5-6.9
Calcium	0.1-10[c]	15.0	0.13	139	0-22
Potassium	0.03-0.7	2.3	0.25	62	854
Bicarbonate	Nil-2.8	58.4	-	-	-
Bromide	0.03	-	-	-	-
Boron	0.01	-	-	-	-

[a]High values found particularly in coastal areas and derived from sea spray.

[b]High values associated with industrial effluent.

[c]High values often over continents and derived from terrestrial dust.

Table 6.3 The chemical composition of water in lake Kivu: the maximum depth 485 metres, anoxic from 70 metres (Degens et. al., 1972)

Depth (metres)	*Ca*	*Mg*	*K*	*Na*	*So⁻₄*	*P*	*Si*	*CO_2*	*NH_3*
	(mg/L)					*(µg/L)*		*(mmol/kg)*	*(µg/L)*
Surface	4.8	87	92	121	23	0.8	231	12	18
100	64	147	145	192	25	18	424	29	487
200	83	182	178	244	166	22	470	24	1314
350	110	394	315	465	214	53	1050	133	5400
410	112	417	338	487	220	54	1226	171	7100

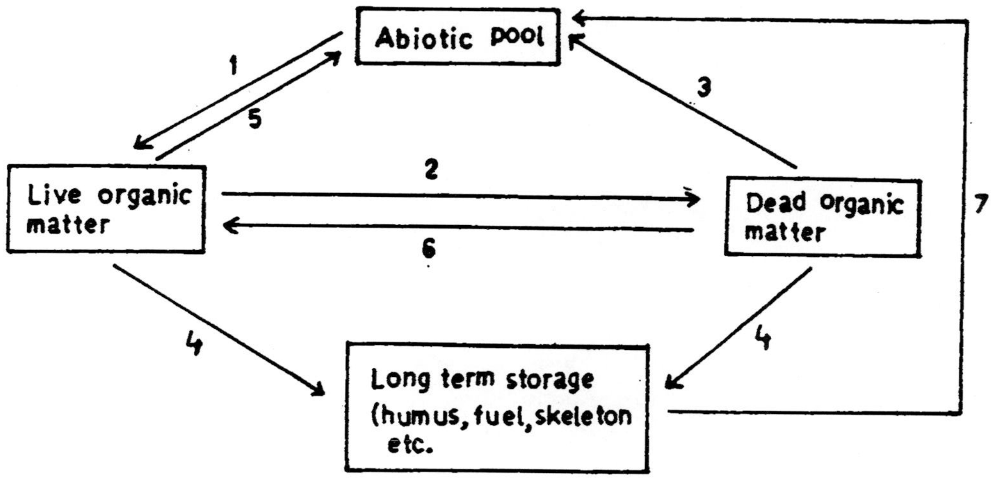

Fig. 6.1: A generalized biochemical cycle. Uptake (1); Death (2); Autolysis, decomposition (3); Sedimentation (4); Excretion (5); Detrivory (6); Slow decomposition, volcanic eruption, burning (7).

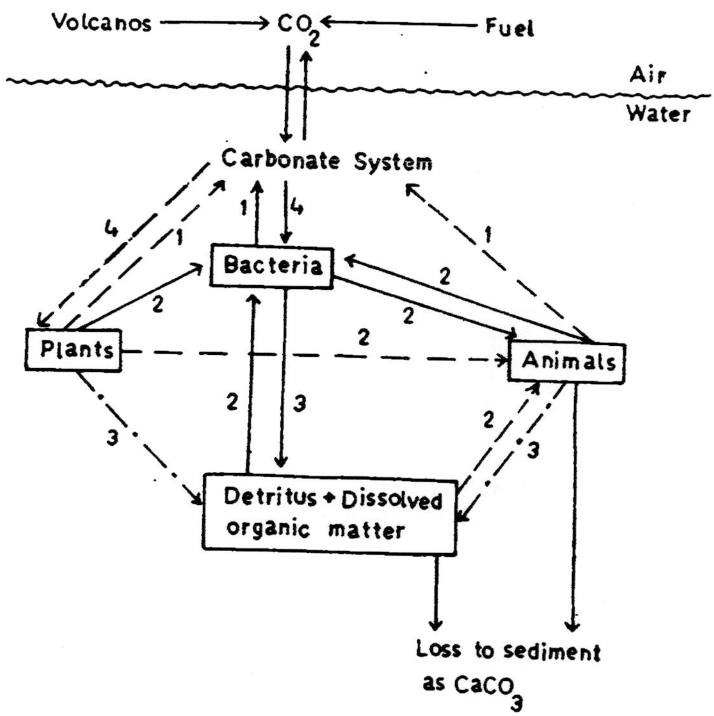

Fig. 6.2: The carbon cycle in lake. Respiration (1); Nutrition (2); Excretion, Decomposition (3); Carbon fixation (4).

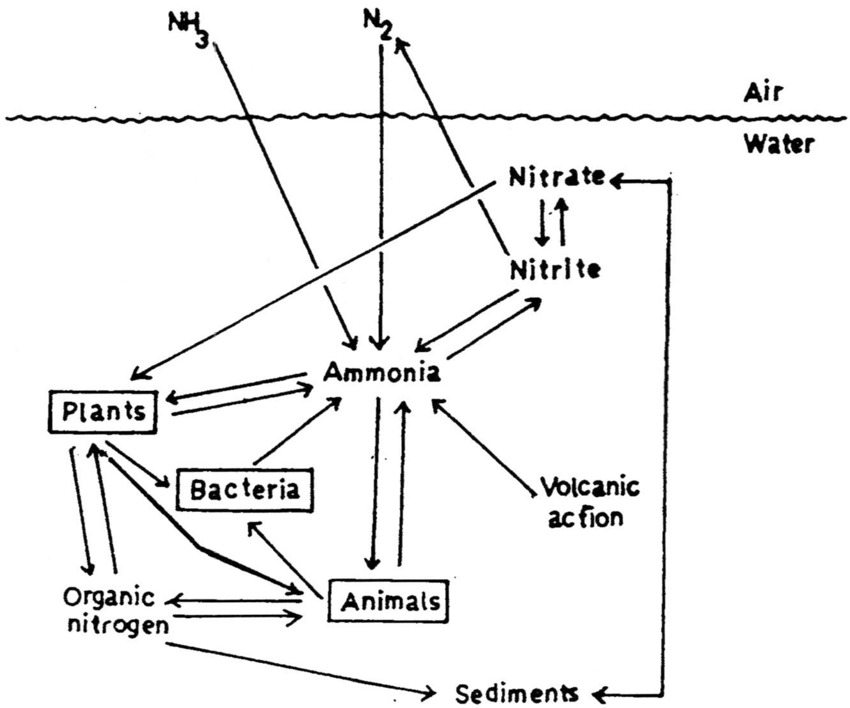

Fig. 6.3: The nitrogen cycle.

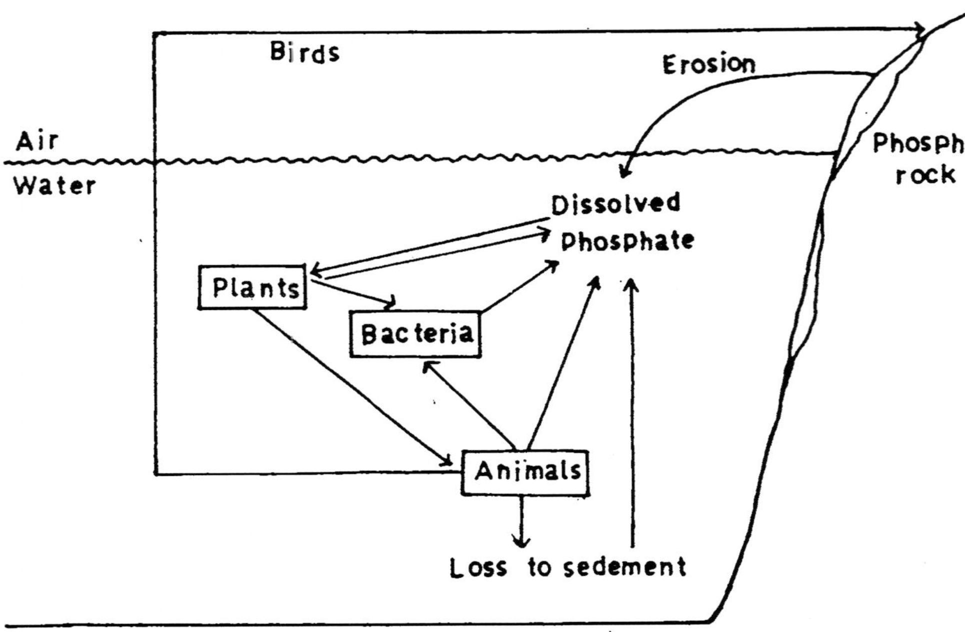

Fig. 6.4: The phosphorus cycle.

VII

Biotic Relationship Among Organisms

The biotic relationship involves the direct or indirect influence of an organism upon the other and may be of following kinds in any lake water.

7.1 SYMBIOSIS

It is an association between two different organisms in which both of them are mutually benefited. This association may be the product of adaptations to long-term environmental alterations during evolution of eukaryotic forms of life on the earth. The following types of symbiotic associations are generally recognized in any lake water.

7.1.1 Alga and animals association

The green alga *Chlorella* is found to be endosymbiont in different animals like *Paramecium bursaria;* sponges *Spongilla, Meyenia* and *Tubella; Hydra* and in some rhabdoceolous tubellarian flat worms residing in freshwaters. All of these animals live in shallow waters where they can obtain adequate light so that the alga can carry on the photosynthesis.

It may well be that the alga is benefited by this association through an access to a larger and more reliable source of nutrients, in the form of metabolic products of animals, than it would has otherwise be possibly for an alga in open water. Thus, the bulk of nitrogen and phosphorus are locked up within this living system, where they are efficiently recycled between an alga and animal. The *Chlorella* also obtain protection from the damaging effect of intense solar radiation of tropics by living within the tissue of animal. The *Chlorella* cells produce sugars like maltose, glucose, etc. which increases the growth rates of animal in under - fed waters. The alga also release oxygen and the animal will obtain additional amount of it, produced by the endosymbiont. The perpetuation of symbiotic relationship depends either upon the engulfment (phagocytosis) of algal cells, a new, from the animal or through transfer of algal cells to next generation of animal via fission, larvae or eggs. The basis for a symbiotic relationship is laid down either when the algal cells are resistant to digestive action of animals or the animals lack the enzymes to digest the algal cells.

The algal cells number in an animal cell is regulated by the following ways : (i) the animal cell may divide in synchrony with algal cells, partitioning equal number of algal cells to animal's daughter cells, (ii) the animal cell may eliminate the excess of algal cells by exocytosis or digestion, or (iii) the animal cell may control the algal cell number by prohibiting the algal cell division through withholding critical nutrients.

Those sponges which are symbiotic with alga increases their light gathering ability by morphological flattening of their body, thereby increasing their surface area available to intercept the light. The digestive regions of the animals also get reduced so that they are less responsive to animal food.

7.1.2 Alga and a higher plant association

The *Azolla* is a small delicate fern which floats freely on the water surfaces of ponds and lakes. It contains a blue-green alga *Anabaena azollae* in the cavities of upper lobes of its floating leaves. The *Anabaena* filaments present in the water comes into contact with leaf primordia in the apical meristem of the *Azolla* and an initial pocket develops on the underside of the leaf primordia to form a cavity which entrap the alga. Within this cavity, the algal cells divide and about 20-50 percent of them will differentiate into heterocysts. The alga is capable of providing all nitrogen requirements of the *Azolla and the Azolla* provides reduced carbon to support nitrogen fixation by the alga. Besides receiving protection from the detrimental effects of low pH of pond's water, the alga does not appear to be in a more favourable position in the *Azolla* than in the free-living form. The alga grow slowly and has decreased rate of photosynthesis inside the host. This symbiotic association appears to be a one sided exploitation of the nitrogen fixing ability of the blue-green alga for the benefit of the *Azolla.* The algal filaments may also be incorporated in the megasporocarp and thus insuring the continuity of the symbiosis in the next generation.

7.1.3 Alga and bacterial association

A bacterium *Pseudomonas fluorescens* is found to be present in the surrounding mucilage of *Volvox aureus* coenobium, and both, the parent and the daughter coenobia of *Volvox* are not found to be viable unless associated with bacterium and this can be due to an extracellular organic factor released by the bacterium. It is probable that the dissolved organic matter leaked from the algal cells are taken up by the bacterium and in this way both of them are benefited. It has been observed that *Oedogonium* filaments rarely produce gametes if grown in axenic cultures, but if they are grown in presence of *Corynebacterium* or *Pseudomonas putrida*, they develop reproductive cells. Similarly, some bacteria are found to be present endosymbiotically inside the nucleus of euglenoids and cytoplasm of dinoflagellates *Gymnodinium, Ceratium* and *Amphidinium.*

The bacterial association with diverse blue-green algal members are commonplace in nature. It has been shown repeatedly that both groups of microorganisms are benefited from each other's presence. The heterocysts of blue-green algae are known to be the sites of organic matter losses and the associated bacteria are chemotactically attracted to a variety of organic carbon (sugars, organic acids etc.) and nitrogen (amino acids etc.) compounds released by the heterocysts. These bacteria in turn locally consume the oxygen during their metabolic utilization and breakdown of excretory products. An associated bacteria can in this manner be considered as a metabolic barrier to the potential oxygen availability for inactivation of nitrogenase activity. The bacterial mineralization of organic matters released by blue-green algae produce carbon dioxide which is immediately utilized for algal photosynthesis and in this manner the associated bacteria can optimized the blue-green algal growth. The other benefits thought to be derived from such associations includes (i) the phycosphere recycling of the phosphorus and nitrogen through bacterial metabolism, (ii) supplementation of growth factors including vitamins by bacterial cells and (iii) the production of metal-specific chelators and surface active compounds by bacterial associate that may play a role in regulating the trace metal availability to alga.

7.2 PARASITISM OF ALGAE

The parasites are organisms living in or upon other organisms and derive nourishment from them. The parasitism involves the penetration of an algal cell by infecting parasitic organisms to obtain food from its cytoplasm. The algal parasitism is not usually uniform and a wide variety of viruses, bacteria and fungi have been isolated from diseased aquatic algae. The ecological effects of parasitism are more or less similar to that of grazing because in both the cases the growth rates of an algal species and the species composition of an algal population are changed and a direct energy flow from one trophic level to another, coupled with the formation of organic detritus from the damaged algal cells is established.

7.2.1 Parasitic virus

The blue-green algae are most commonly parasitized by viruses, but *Chlorella cells and Oedogonium* germlings are also parasitized by them. The parasitic viruses on algae are virtually identical with those infecting bacteria. The tail of viruses attaches to an algal cell wall and the viral DNA enters into the host cytoplasm via the tail and will multiply in the host cytoplasm causing lysis of the cell and the liberation of new virus particles. The resistant strains of blue-green algae have also been reported. Since, blue-green algae are undesirable contaminants of water reservoirs, swimming pools etc., it has been suggested that some measure of control is possible by adding virus cultures to the waters.

7.2.2 Parasitic bacteria

Like viruses, the parasitic bacteria tend to be more abundant on blue-green algae and can cause their lysis. The heterocysts are usually immune to bacterial attack. Since, blue-green algae have very little or no cellulose in their cell walls, they are less protected from bacterial attack than other groups of algae having cellulosic walls (bacteria degrade cellulosic wall less easily). The parasitic bacteria are usually pigmented, rod shape, aerobic, gram negative and non-flagellated.

7.2.3 Parasitic fungi

The fungi of groups Chytridiomycete (chytrids) and Phycomycete (usually the biflagellate holocarpic forms) are main fungal parasites of algae. The chytrids are mainly external parasites, while phycomycetes are usually internal parasites. The chytrids infection is usually brought about by zoospores, which attaches to the algal cell walls or mucilage and produces germ tubes, which penetrates into algal cells. There is a well - marked host specificity of some of these fungal parasites, e.g., dinoflagellates *Ceratium* and *Peridinium* are parasitized by chytrid *Amphicypellus elegans,* the cysts of *Ceratium* are parasitized by chytrid *Rhizophidium nobile,* the diatoms *Fragilaria crotonensis* and *Asterionella formosa* are parasitized by *Rhizophidium planktonicum,* and green algae *Scenedesmus* and *Pediastrum* are parasitized by *Phyctidium* sp.

The phycomycetes *Saprolegnia, Aphanomyces* and *Pythium* spp can parasitize the *Spirogyra* filaments. The vegetative hyphae of *Saprolegnia asterophora* can invade the healthy algal filaments from the infected algal filaments and can cause the infection. The fungal infection often distort the spiral arrangement of chloroplasts and disintegrate the pyrenoids of the host alga.

The relative movement of algal cells and the fungal spores in turbulent waters makes their contact a possibility and from this it can be seen that why fungal epidemics are more common in a densely populated lakes. In all cases, the number of algal cells must increase significantly in order to be parasitized. In fact, apparently, very healthy algal cells are readily parasitized, but during their fast growing periods only few epidemics seems to take place because algal cells recover from extensive parasitism by rapid division of a small number of healthy algal cells to establish a population.

7.3 PREDATION OR GRAZING OF ALGAE

The predation or grazing involves the consumption of algae by different species of animals and is a powerful biotic force in controlling the species composition and biomass of algal communities. In freshwaters, the algae are subject to grazing pressure principally from protozoa, rotifers, crustaceans and fishes. Rotifer *Asplanchna* selectively graze *Staurastrum,* while crustacea prefer to eat chysophytes than diatoms or green algae. Snapping turtles eat *Spirogyra crassa* and *Cladophora* sp., and aquatic insect *Leptophlebia vespertina* feed on desmids, diatoms and filamentous green algae. The man, an active predator of herbivorous fishes, turtles etc., can also alter the natural relationships between plants and animals.

The grazing of algae commonly reduces their photosynthetic surface area, shorten their life span, increases the mortality of their germlings and reduces their population size. The phytoplanktons are able to replenish very quickly after being grazed because of their rapid rate of reproduction. The small size of most phytoplanktons makes them more suitable as a food for a much wider variety of herbivores in the food chain than higher plants. The herbivores usually consume about half or more of their body weight per day, but the algae reproduces usually at the rate of 1-1.5 divisions per day. An increase in water turbulence usually reduces the feeding rate of animals. The impact of grazing on algae varies with seasons and is found to be maximum in summer when zooplanktons are more active than in other seasons.

7.3.1 Impact of grazing

(i) Summer minima of algal populations in temperate lakes

The grazers are presumably present in the water all the time, but the chances for them to find their prey are small except during algal bloom conditions. The grazing may be a factor in maintaining the summer minima of algal populations in temperate waters and it is for this reason that the population density or the standing crop of the phytoplanktons may be a poor or misleading indicator of primary productivity. The major part of the primary productivity resides not in the bodies of the plants but in that of zooplanktons. Since the breeding cycle of grazers is much longer than the reproductive cycle of phytoplanktons, the grazers can not build up a large population until the phytoplanktons have not bloomed.

A small summer bloom of green algae in many lakes has never been adequately explained, but it has been considered that an important factor responsible for such a bloom may be nutrients regeneration by zooplanktons.

(ii) Reduction of competition among algae

The grazing avoid the build up of heavy phytoplankton populations and thus improve the chances of co-existence among them. The grazing decreases the total algal biomass and thus reduces the demand on nutrients and also the severity of competitive interactions among algae.

Under high grazing pressures, all but the most resistant species are consumed; while under low grazing pressure a competitive dominant will occupy all the available space. Therefore, the algal diversity is maximum at intermediate levels of grazing.

(iii) Algal size and grazing

Algal size and shape are important factors as far as grazing of algae by animals is concerned. It is frequently reported that small cells in an algal population are easily grazed then large cells, filaments or colony. Although no size class or group of algae is immune to grazing, many desmids, diatoms and dinoflagellates are too large or unsuitably constructed to be grazed by zooplanktons. Large dinoflagellates like *Peridinium* and *Ceratium* are not at all grazed by filter - feeding zooplanktons and are grazed only by rotifers *Hydatina* and *Asplanchna*. Cladocerans such as *Daphnia* can change the composition of phytoplanktons from nanoplankton dominance to net plankton dominance, because they have difficulty to feed large forms. Despite the fact that large algal cells, filaments or colony are not easily grazed by crustaceans, many fishes and birds have developed several means to utilize them as food. The dinoflagellate *Peridinium* and several colonial blue-green algae are easily grazed and digested by cichlid fish *Tilapia galilaea*.

The grazing by filter - feeding animals is generally size selective with the upper threshold of potentially edible algal cells increases as the body size of animal increases. The zooplankton grazing can therefore drastically alter the size distribution and species composition of phytoplankton populations and the result being highly dependent on the composition of zooplanktons assemblage themselves.

(iv) Patchiness of phytoplankton

The distribution of phytoplankton is uneven, and one of the major factor responsible for this uneveness or patchiness is uneveness of the distribution of grazers themselves. There are patches of intensive feeding in which almost all algae are eliminated, while nearby there are algal patches which are unattacked. The grazers cause the patchiness of the phytoplanktons rather than phytoplanktons exclude the grazers. The patchiness of phytoplankton is essential to allow the cells to grow until the patch is discovered again by the grazers (Fig. 7.1).

(v) Remineralization of inorganic nutrients

In addition to allowing the organic materials to enter the higher trophic levels, the grazing is first step in the remineralization of inorganic nutrients to support the growth of other algae. The break up of algal cells during passage through animals gut release nutrients which are excreted back into water. Under these circumstances, nutrients would be constantly regenerated in the upper water layers and the productivity of algae will never by halted. Under optimal feeding conditions, consumers excrete nitrogen equivalent to about one-third ingested by them and there exist an intimate link between grazing, nutrients recycling and phytoplankton growth.

(vi) Provision of nutrients to algae

The grazing may actually be beneficial to some algae. It has been reported that the colonies of green alga *Sphaerocystis schroeteri* are only partially disrupted while passing through the gut of *Daphnia magna* and most of them emerge intact and in viable conditions. During passage through the gut, the alga take up phosphate and other nutrients from the gut fluid and which will stimulate the photosynthesis and cell division in defecated alga. The enhanced algal growth will compensate much more than minor losses incurred due to grazing.

(vii) Toxicity to animals

Some dinoflagellates like *Ceratium, Gymnodinium* etc. contains toxic substances and the predators which consume them, either directly or indirectly, will become ill or die due to food chain amplification of toxins. The degree to which they are affected depends upon the dinoflagellates population size, the rate of feeding, the duration of exposure while feeding and the storage phenomena. The toxins are initially concentrated in the liver but can spread throughout the body. The poisoning is accompanied by the gastrointestinal complaints, vomiting, abdominal pain and diarrhoea. Similarly, a poisonous species of blue-green alga *Lyngbya* is detected in the alimentary tract of large number of fishes.

7.3.2 Adaptations in algae against grazing

(i) Spiny outgrowth

Some algae like desmids develops surface outgrowth or protuberance, which can reduce chances of cells being eaten by animals and can be interpreted as anti-predatory device. The seta of *Coleochaete* defend the alga against grazers, as broken seta exude a substance which repel potential grazers. The enclosure of cells within the layer of siliceous scales with elongate bristles (as in some chrysophytes members) may also prevent grazing. The organic scales which are spiny or in any way raised up from the cell surface may limit the chances of predation. The large robust horns of *Ceratium* prevents its entry into oral aperture of small animals.

Similarly, many algae with gelatinous sheath seems to be protected from grazing. The cryptomonads may embed themselves in palmelloid mucilage layers and which may be effective against some grazers.

(ii) Large body size of algae

The large body size of algae, as can aggregation of cells into filaments or colonies can help them to escape the influence of smaller grazers. Many diatoms are too big or form chains or star - shaped colonies or develop spines, which may prevent them to be grazed easily. Similarly, due to relatively large cell size, most of the dinoflagellates can not be used directly as a food by freshwater grazing animals. The colonial green flagellates, like *Volvox,* escape grazing due to their comparatively large body size and are usually present in the water when grazing pressure from filter-feeding zooplanktons is likely to be high.

(iii) Rapid growth of algae

It is probable that the most common and effective mean of avoiding grazing pressure by phytoplanktons is their rapid rate of growth and reproduction.

(iv) Crustose morphology of algae

It has been suggested that the heteromorphic life cycles in algae involving a crustose thallus may have evolved in response to grazing pressure. The grazing difficulty increased from unicellular through filamentous, foliose and ultimately to crustose alga.

(v) Calcification

The calcification of some algae, like *Chara*, may be of survival value under high grazing pressure.

(vi) Algal movement

The swimming by flagellated algae may be helpful to avoid predation.

(vii) Toxicity in algae

The presence of toxic factors in algae may protect them against grazing. The endo- and exotoxins of some blue-green algae like *Anabaena flos-aquae, Microcystis aeruginosa* and *Aphanizomenon flos-aquae* (mostly of peptides and alkaloids in nature, Codd and Poon, 1988) may reach levels sufficient to kill some zooplanktons and fishes and may be partly responsible for exclusion of grazers. The feeding behaviour of cladocera is often negatively affected by bad taste, restricted ingestion and assimilation of blue-green algae. Similarly, the release of toxic metabolites by certain dinoflagellates bloom may discourage zooplanktons grazing.

7. 4 COMPETITION AMONG ALGAE

The competition is an ecological term which refers to an interaction among different organisms for a necessary resource such as light, nutrients or space existing in short supply. The coexistence of different organisms competing for a resource can occur if they are spatially separated, e.g. if the one is benthic and the other is planktonic. The two diatoms *Achnanthes lanceolata* and *Cocconeis placentula,* both growing adnate to a substrate never equally dominate at the same time. Although, the total space available to an alga may not be usually limiting, but there may only be a limited number of suitable microhabitats (supplied with optimum levels of light, nutrients etc.) available for colonization of an alga. There is a competition for usable rather than total space. The competition is observed when an alga overgrow or shade the other. The benthic space supplied with adequate available light is the principal limiting factor for the growth of benthic algae.

In a competitive interaction, the competitors may either manage to share the limited resources or may exclude one another. In the first case, both of them are hampered with respect to their growth and reproduction, while in the second case one of them is totally eliminated. The heavy settlement of an algal spores on a substratum may lead to competition for light, nutrients, space etc. between germlings and an inverse relationship exist between population density and individual size. It is also possible than an algal species best suited to an environment may exclude all others from that environment and this is known as competitive exclusion.

The competition may either be among individuals of the same species (intraspecific) or between two closely related different species (interspecific). The competition has various possible outcomes. It lay the development of a three dimensional community and maintain a high species diversity.

A successful competitor may have either one or all of the following characters : (i) it should tolerate extreme environmental conditions so that it can grow and reproduce on a wide range of substrates and under variety of conditions, (ii) it should have a rapid rate of growth and reproduction and should disseminate widely. A rapidly growing algal species is able to outcompete a slowly growing algal species, (iii) it should have some ability to produce toxins, antibiotics etc. and at the same time must be able to tolerate the toxins of other organisms, so that it can colonize a micro-habitat, and (iv) it should have some ability to avoid competition by living in habitats not occupied by other organisms (e.g. the existence of some blue-green algae in salt lakes or thermal springs).

7.5 AUTO AND ALLO-INHIBITION AND STIMULATION AMONG ALGAE

There have been intermittent reports of an inhibitory or a stimulatory effect of an alga upon itself (autoinhibition or auto-stimulation) or upon others (alloinhibition or antibiosis or allostimulation) due to release of several extrametabolites (Hellebust, 1974). An autoantibiosis has been observed in old cultures of diatom *Nitzschia palea.* Such effects are found to be operative only during dense phytoplankton blooms or in small bodies of water e.g. in ponds or in still water and can be reversed or diluted in fast flowing or turbulent situations.

An extracellular metabolite of an alga may be either inhibitory or stimulatory to one algal species but not to others and thus may determine the distribution of algae in nature. In mixed cultures of *Chlorella vulgaris* and *Nitzschia frustulum*, mutual antibiosis seems to occur. Similarly, *Asterionella formosa* and *Nitzschia palea,* when grown together produce antibiotics against each others. An interdependent association of two species is only possible if their extrametabolites are not toxic to each other. For example, seldom if ever the blue-green algae *Microcystis aeruginosa* and *Aphanizomenon flos-aquae* are found growing in the same body of water at the same time, although the habitat may be highly suitable for both of them, due to their mutual antibiosis. In contrast, in hard water lakes adequately supplied with nutrients, there is almost invariably an association of *Lyngbya birgei, L. major, Coelosphaerium kuetzingianum, Microcystis aeruginosa, Asterionella formosa, Pediastrum boryanum and Scenedesmus quadricauda.*

7.6 IMPACT OF HUMAN BEINGS ON ALGAE

Human beings are responsible for various kinds of pollution in any freshwater ecosystem. Anything which results into changes in the physical, chemical or biological properties (like changes in the balance of species, reduction of species diversity or dominance of only few species) of an ecosystem is called a pollutant. A lake water may have following types of pollution.

7.6.1 Eutrophication

Actually it is not a pollution but an enrichment of lake water with nutrients either by natural (in long course of time) or artificial (man made, in short period of time) means. In general, a deep lake can withstand more nutrients input than a shallow lake without an adverse effect. A continuous enrichment of lake water with nutrients such as nitrogen, phosphorus and carbon may occur by the addition of organic substances derived either from household sewage, phosphate containing detergents, drainage from agricultural lands, waste from milk plants, canneries, paper mill, starch factories and fish processing plants or their decomposition products.

Due to eutrophication in a lake, the rate of gross primary production may increases from 0.5 to 5.0 grams dry organic matter/m^2/day during most favourable growing season. The eutrophic lakes largely supports the growth of few phytoplanktons like blue-green algae *Microcystis, Anabaena, Phormidium, Oscillatoria, Spirulina, Lyngbya* etc.; diatoms *Asterionella, Fragilaria, Cyclotella, Melosira, Tabellaria etc.;* green algae *Chlorella, Ankistrodesmus, Scenedesmus, Eudorina, Stigeoclonium* etc.; and some euglenoids. The nitrogen fixers among blue-green algae can further increase the nitrogen content of lake water, if other nutrients (particularly the phosphorus) are supplied in excess and thus confirming the prime role of phosphorus for eutrophication.

The floating mats of green and blue-green algae on the water surface may prevent the passage of light into water and can inhibit the photosynthesis in planktonic algae and submerged plants and can produce oxygen starvation in water and the subsequent death of fishes and other animals. They also interfer with fishing, boating, swimming and fish culture and some of them may produce toxic substances against animals. The oxygen released by the surface mats of algae largely escape into air and when large number of algae die, they requires large quantities of dissolved oxygen for their decomposition and will produce carbon dioxide and ammonia (which in higher concentrations may prove toxic to fishes). The amount of dissolved oxygen required to degrade the dead organic matter until the water is again purified is its biological oxygen demand (B.O.D.).

The eutrophication can be checked to some extent either by (a) removal of undissolved and dissolved nutrients from the waste water or sewage effluents through primary, secondary and tertiary processes (the dissolved phosphorus in the water can be removed through chemical precipitation by addition of ferric chloride), (b) controlling the growth and multiplication of algae and higher plants by application of suitable doses of copper sulphate and sodium arsenate, respectively or (c) by harvesting and mechanical removal of algae, higher plants and fishes, which will subsequently reduce the amounts of nutrients recycled into water after their death.

7.6.2 Toxic chemicals

Various toxic chemicals produced by different industries find their way into lakes. These chemicals are not usually metabolized and get accumulates in the body of an organism and when several such individuals are consumed by a carnivore of the next trophic level, the total concentration of that chemical will increase many times in its body. The continuation of this process will tend to increase the concentration of that chemical several times in the top carnivore if the food chain is long. The human beings may also be severely affected because they consume organisms from the higher trophic levels. A continuous increase in the concentrations of a chemical in the bodies of organisms of succeeding trophic levels is called the *Biological Magnification.*

(i) Heavy Metals

The heavy metals like copper, chromium, iron, mercury etc. when enters into a water body decreases the diversity and productivity of algae and when such metals are added in combination they will produce a more pronounce effect than when added separately. For example, when copper and zinc are added together they will act synergistically and will produce a greater effect than sum of these chemicals separately.

Some algae can tolerate heavy metals either by changing their membrane permeability, internal detoxification of heavy metals or extracellular chelation of them. The permeability of a cell membrane can change in response to a heavy metal so that a metal is prevented from reaching the site of its action. The tolerance to heavy metals appears to be mainly due to exclusion rather than detoxification as non-tolerant strains accumulate several times of copper than a tolerant strain, but a number of tolerant benthic diatom species have been found to absorb and detoxify copper by localizing it in the cell as insoluble polyphosphate - copper bodies.

The copper is toxic to many algae at comparatively low levels (0.30 mg/L), but in alkaline waters it will precipitate out quickly as copper carbonate and more slowly as copper hydrate and will be effective as an algicide only for a short period following its application. A number of algae like *Nostoc, Calothrix, Chara, Nitella, Pithophora, Eudorina, Pandorina, Scendesmus, Stigeoclonium, Cymbella, Navicula, Asterionella* are found to be very resistant to toxic effects of copper sulphate.

Some fungicides contain mercury compounds and the increasing levels of these compounds in a water body will depress photosynthesis in algae. The methyl mercury chloride may accumulate upto dangerous levels in the bodies of freshwater fishes, birds and human beings via food chain amplification.

(ii) Pesticides

Pesticides like Dichlorodiphenyl - trichloroethane (DDT), Dieldrin etc., when applied over an agricultural land may enter into a water body via land drainage and produce serious effects to the biota. They are relatively insoluble in water but are more soluble in body fats or lipids. They are remarkably stable compounds in the ecosystem and adheres strongly to the solid particles and upon the algal surfaces and undergo biological magnification through succeeding trophic levels to the top carnivore level. They produce serious effects to freshwater birds because they interfere with calcium deposition in egg shell and as a result eggs with thin shells are produced which break upon when birds incubate them.

(iii) Oil

The oil wells and refineries release oil, oily soda, tar, lime water, sludge ammonia and phenol and when these substances enters into a water body, they either floats on the water surface or sink to the bottom and may destroy aquatic organisms. The toxic components present in the crude oil are volatile aromatic compounds like benzene, toluene, xylene etc. The oil usually forms a film on the water surface and can interfere with gas exchange between air and water. When birds come in contact with floating oil, they will be coated with oil and their feathers will loss the insulating properties and as a result most of them will be unable to feed and will die. The immediate effect of oil is the loss of herbivores and a subsequent increase in algal flora, but thick oil deposits may smoother algae and restrict the recolonization of area by algae. The oil disrupt the cell membrane and restrict the diffusion of inorganic carbon into algal cells.

However, some diatoms like *Melosira varians, Amphora ovalis, Navicula radiosa, Synedra acus, Asterionella* sp., *Cyclotella* sp. and *Fragilaria* sp. can resist oil pollution to some extent.

7.6.3 Powdered particles

They are the particles of soil, sand, silt, coal, mining waste etc. conveyed into water from land drainage. They may either remain suspended in the water for some time or may settle quickly to the bottom. When they are remain suspended in the water, even for some time, they increases the turbidity of water, reduces the light penetration and suppress the growth of algae. A depression of carbon fixation and a temporary cessation of algal growth occurs at low light intensities. But when they settle to the sediment, they will suppress the growth of benthic algae and aquatic plants, destroy the bottom-dwelling organisms and reduces the food availability to fishes and also the capacity of the lake.

7.6.4 Radioactive waste

The major sources of radioactivity in any water body have been the fallout from nuclear weapon testing, dumping off of waste from nuclear fuel cycle system mainly from power or reprocessing plants, mining and processing of radioactive ores and use of refined radio-active materials for industrial, medical and research purposes.

An alga can take up an element either through engulfment or absorption/adsorption on to exposed surface area and can assimilate and store it irrespective of that that whether it is a radioactive element or not. Some radioactive elements may accumulate in an organism to an extent of many thousand times than their concentrations present in the water. The *Spirogyra* cells can concentrate radiophosphorus upto an extent of 850,000 times to that present in the water. It is evident from Table 7.1 that the organisms of lower trophic levels (e.g. algae and insect larvae etc.) can concentrate the radioactive substances much more than those of higher trophic levels (e.g. fishes). Probably the fishes tend to eliminate the non-essential elements making a selection against such radio-isotopes along the food chain.

The ionizing radiations disrupt the chemical bonds in the molecules of living cells and produce both somatic and genetic damages. Generally, the microorganisms and lower plants have been found to be more resistant to a particular dose of a radioactive substance than higher plants or animals.

7.6.5 Acid pollution

The drainage from mining sites or from some other industries can produce extremely acidic waters of pH below 3. The sulphur dioxide emitted from combustion of fossil fuels and supplemented with that derived from volcanos, and the hydrogen sulphide formed by bacterial activities on the excessive amounts of organic matters present in many shallow water bodies may enter into a lake either directly or as acid rain and can lower the pH of its water. The low pH water contains excessive amounts of metals like aluminium, manganese, zinc, iron etc., which are leached from catchment areas and lake sediments. Thus the algae growing in acidic waters are also more resistant to heavy metals.

The volcanic lakes, bog pools or acid polluted waters have pH usually less than 4.0. At pH below 5.0, only free carbon dioxide is of any quantitative importance to algae. The acidification of waters causes the reduction in species diversity and density. Those algae which are able to colonize such waters are able to exclude excess hydrogen ions. The members most commonly recorded in acidic waters are some green algae like *Chlamydomonas acidophila* and *C. applanata;* dinoflagellates like *Gymnodinium uberrimum, Peridinium inconspicuum, P. pusillum, P. willei;* diatoms like *Frustulia rhomboides, Eunotia exigua, E. lunaris, E. pectinalis, Navicula minima, N. viridis, N. subtilissima, Pinnularia viridis* etc.; chrysophytes like *Mallomonas, Synura, Chromulina, Ochromonas, Uroglena* etc. and *Euglena mutabilis.*

7.6.6 Thermal Pollution

The thermal power stations and atomic reactors use large quantities of waters for cooling purpose and release lot of hot water in nearby lake or river and thus can increase the temperature of a lake water. When the temperature of a lake water is maintained at above 35°C, the usual flora of diatoms and green algae are replaced by excessive growth of few tolerant blue-green algae such as *Synechococcus, Lyngbya, Mastigocladus, Oscillatoria, Phormidium, Plectonema* and *Scytonema*. Some of these algae can tolerate a temperature increase from 40 to 85°C. A seasonal sequence of algae has also been observed to occur in the same direction (diatoms ---> green algae ---> blue-green algae) in a temperate lake as its water warms up, as with the onset of thermal pollution.

With the rise in temperature of water, the solubility of oxygen decreases and that of salts increases. The decay of organic matter and the rate of respiration, both increases as the water temperature increases. Hence, the biological oxygen demand of warm water is much more than that of cold water. The hot water may kill the fishes and other aquatic animals directly and can increases the susceptibility of organisms to toxic materials. The ambient temperature of a lake water is also very important as far as the severity of a hot water effluent is concerned. The deleterious effect of a thermal discharge is more pronounce in the tropics, where the organisms are already living at temperatures close to their thermal maximum and a little increase in temperature is sufficient to stress the tropical lakes than a temperate, an arctic, antarctic or an alpine lake.

Table 7.1 : The maximum concentration factors for different radio-isotopes in Columbia river organisms

Isotopes	*Algae*	*Insect larvae*	*Fish*
p^{32}	1000,000	100,000	100,000
Zn^{65}	100,000	10,000	10,000
Na^{24}	100	100	1000
Cu^{64}	10,000	1000	10
Cs^{137}	5000	1000	10,000

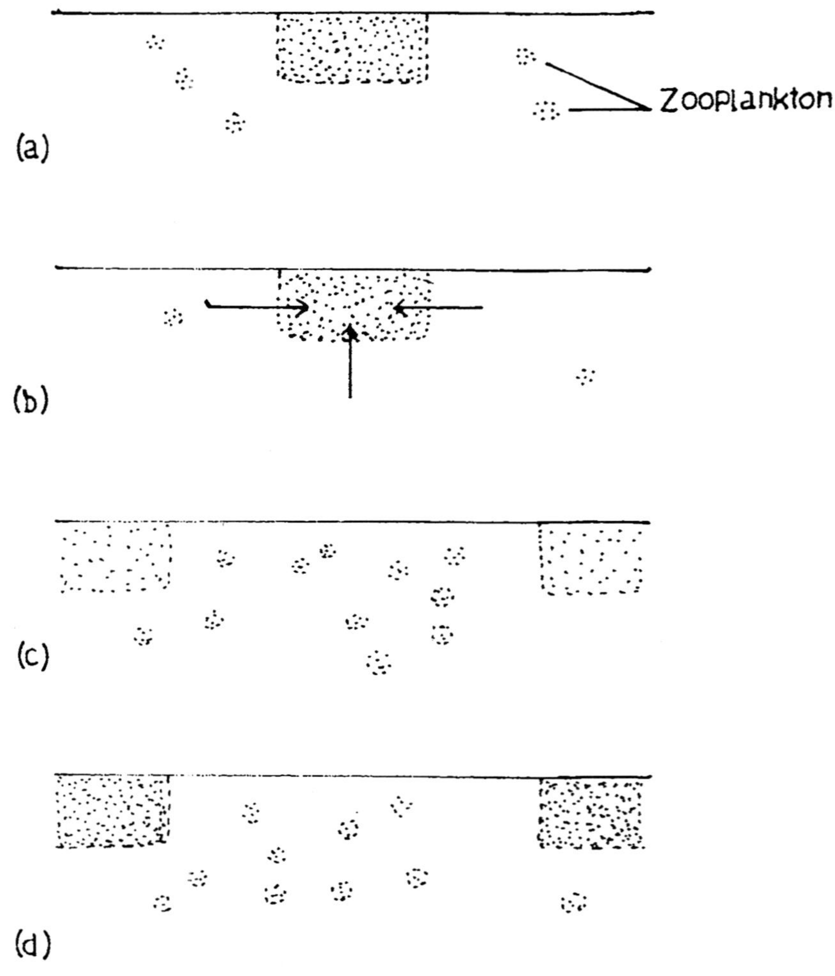

Fig. 7.1 : Diagrammatic summary of phytoplankton - zooplankton interrelationships. (a) Inverse relationship at start, (b) Commencement of zooplankton migration with some grazing, (c) Heavy grazing on completion of migration, growth of phytoplankton on either side, (d) Return to inverse relationship. Density of phytoplankton is indicated by degree of strippling

Bibliography

Behre, K. (1956). Die Algenbesiedlung Seen um Bremen and Bremerhaven. *Veroff. Inst. Meersforsch. Bremer haven* 4, 221-383.

Codd, G.A. and Poon, G.K. (1988). Cyanobacterial toxins. In *Biochemistry of the algae and cyanobacteria* (eds. L.J. Rogers and J.R. Gallon), pp. 283-296. Clarendon Press, Oxford.

Degens, E.T., Okada, H., Honjo, S. and Hathaway, J.C. (1972). *Micro-crystalline sphalerite in resin globules suspended in lake Kivu, East Africa Miner. Depos.* 7, 1-12.

Engelmann, T.W. (1882). Über Licht-und Farbenperzeption niederster organismen. *Pflügers Arch. f. d. ges. Physiol. d. Menchen u.d. Tiere.* 29, 387-400.

Goldman, C.R. (1967). Integration of field and laboratory experiments in productivity studies. In *Estuaries* (ed. G.H. Lauff), *Amer. Assoc. Adv. Sci*, Washington D.C., 346-352.

Gorham, E. (1955). On the acidity and salinity of rain. *Geochimica et Cosmochimica Acta* 7, 231-239.

Hellebust, J.A. (1974). Extracellular products. In *Algal physiology and biochemistry* (ed. W.D.P. Stewart), pp. 838-863. Blackwell Scientific Publications, Oxford.

Hensen, V. (1887). Über die Bestimmung des Planktons oder des in Meere treibenden Materials an Pflanzen und Thieren. *Ber. Kommn. wiss. unters. deutsch. Meere. Kiel.* 12-14, 1-107.

Hickel, B. (1978). Phytoplankton population dynamics in PluBsee (East Holstein, Germany). *Verh. Ges. Okol. Kiel.* 1977 (1978), 119-126.

Hickel, B. (1985). The population structure of *Ceratium* in a small eutrophic lake. *Verh. Internat. Verein. Limnol.* 22, 2845-2849.

Hutchinson, G.E. (1957). *A treatise on limnology.* Vol. I. *Geography, Physics and Chemistry.* pp. 1-1015. J. Wiley & Sons, New York.

Hutchinson, G.E. (1967). *A treatise on limnology.* Vol. II. *Introduction to lake biology and the limnoplankton.* pp. 1-1115. J. Willey & Sons, New York.

Juday, C. (1942). The summer standing crop of plants and animals in four Wisconsin lakes. *Trans. Wisconsin. Acad. Sci.* 34, 103-135.

Livingston, D.A. (1963). Chemical composition of rivers and lakes. *Prof. Pap. U.S. geol. Surv., no.* 440, G1-G64.

Nygaard, G. (1949). Hydrobiological studies on some Danish ponds and lakes, part II. The Quotient hypothesis and some new or little known phytoplankton organisms. *Kgl. Dansk. Vidensk. Selskab.* 7, 1-293.

Ryther, J.H. (1956). Photosynthesis in the ocean as a function of light intensity. *Limnol. Oceanogr.* 1, 61-70.

Sieburth, J.M.N., Smetacek, V. and Lenz, J. (1978). Pelagic ecosystem structure : Heterotrophic compartments of the plankton and their relationships to plankton size fractions. *Limnol. Oceanogr.* 23, 1256-1263.

Sournia, A. (ed.) (1978). *Phytoplankton Manual.* Monogr. on Oceanogr. Methodology no. 6, UNESCO, Paris.

Stull, E.A., Amezaga, E.de and Goldman, C.R. (1973). The contribution of individual species of algae to primary productivity of Castle lake, California. *Verh. Int. Ver. Limnol.* 18, 1776-1783.

Visser, S.A. and Villeneuve, J.P. (1975). Similarities and differences in the chemical composition of waters from West, Central and East Africa. *Verh. Int. Ver. Limnol.* 19, 1416-1428.

SUGGESTIONS FOR FURTHER READINGS

Bold, H.C. and Wynne, M.J. (1978). *Introduction to the algae. Structure and reproduction.* Prentice Hall Inc., Englewood Cliffs, New Jersey.

Boney, A.D. (197' *Phytoplankton.* Edward Arnold, London.

Campbell, R. (1983). *Microbial Ecology.* Blackwell Scientific Publications, Oxford.

Carmichael, W.W. (1981). *Water Environment, Algal Toxins and Health.* Environmental Science Research Vol. 20: International conference on Toxic algae. Plenum Press, New York.

Carr, N.G. and Whitton, B.A. (1982). *The biology of Cyanobacteria.* Blackwell Scientific Publications, Oxford.

Chapman, V.J. and Chapman, D.J. (1973). *The Algae.* MacMillan, London.

Cook, C.D.K., (1975). *Water plants of the world.* Dr. W. Junk b.v., Publishers, the Hague, Netherlands.

Fenchel, T. and Blackburn, T.H. (1979). *Bacteria and Mineral cycling.* Academic Press, London.

Fogg, G.E. (1965). *Algal cultures and phytoplankton ecology.* Athlone Press, London.

Fogg, G.E. (1971). Extracellular products of algae in freshwater. *Arch. Hydrobiol. Beih, Ergebn. Limnol.* 5, 1-24.

Fogg, G.E., Stewart, W.D.P., Fay, P. and Walsby, A.E. (1973). *The blue-green algae.* Academic Press, London.

Fritsch, F.E. (1935). *The Structure and Reproduction of the Algae.* Vol. I. Cambridge University Press, London.

Fritsch, F.E. (1945). *The Structure and Reproduction of the Aglae.* Vol. II, Cambridge University Press, London.

Fryxell, G.A. (1983). *Survival strategies of the Algae.* Cambridge University Press, London.

Golterman, H.L. (1960). Studies on the cycle of elements in fresh water. *Act. Bot. Neerl.* 9, 1-58.

Golterman, H.L. (1975) *Physiological Limnology, an approach to the physiology of lake ecosystems.* Elsevier, Amsterdam.

Gorham, P.R. (1964). Toxic algae. In *Algae and Man* (ed. D.F. Jackson), pp. 307-336. Plenum Press, New York.

Higgins, I.J. and Burns, R.G. (1975). *The Chemistry and Microbiology of Pollution.* Academic Press, London.

Hutchinson, G.E. (1975). *A Treatise on Limnology.* Vol. III, Limnological Botany. John Wiley and Sons, New York.

Jackson, D.F. (1964). *Algae and Man.* Plenum Press,. New York.

Lund, J.W.G. (1959). Buoyancy in relation to the ecology of the freshwater phytoplankton. *Br. Phycol. Bull.* 1 (7), 1-17.

Lund, J.W.G. (1965). The ecology of the freshwater phytoplankton. *Biological Review* 40, 231-293.

Macan, T.T. (1970). *Biological studies of the English Lakes.* Longmans, London.

Morris, I. (1980). *The Physiological Ecology of Phytoplankton.* Blackwell Scientific Publications, Oxford.

Moss, B. (1980). *Ecology of Freshwates.* Blackwell Scientific Publicatons, Oxford.

Odum, E. (1971). *Fundamentals of Ecology.* Saunders, Philadelphia.

O'Kelley, J.C. (1974). Inorganic nuntrients. In *Algal Physiology and Biochemistry* (ed. W.D.P. Stewart), pp. 610-634. Blackwell Scientific Publications, Oxford.

Philipson, J.O.T. (1966). *Ecological Energetics.* Institute of Biology: Studies in Biology I. 'Edward Arnold, London.

Pickett-Heaps, J.D. (1975). *Green Algae.* Sinauer Association, Sunderland, Mass.

Prescott, G.W. (1968). *The Algae: A Review.* Houghton-Mifflin Co. Boston.

Reynolds, C.S. (1984). *The Ecology of Freshwater phytoplankton.* Cambridge University Press, Cambridge.

Reynolds, C.S. and Walsby, A.E. (1975). Water blooms. *Biol. Rev.* 50, 437-481.

Rheinheimer, G. (1980). *Aquatic Microbiology.* John Wiley and Sons, Chichester & London.

Round, F.E. (1973). *The Biology of the Algae.* Edward Arnold, London.

Round F.E. (1981). *The ecology of algae.* Cambridge University Press, Cambridge.

Ruttner, F. (1953). *Fundamentals of Limnology.* D.G. Frey and F.E.J. Fry (Translators). University of Toronto Press, Toronto.

Sculthorpe, C.D. (1967). *The Biology of Aquatic Plants.* St. Martins Press, New York.

Smith, G.M. (1950). *Freshwater Algae of the United States.* McGraw Hill, New York.

Sprent, J.I. (1979). *Biology of Nitrogen - fixing organisms.* McGraw Hill, New York.

Stewart, W.D.P. (1974). *Algal Physiology and Biochemistry.* University California Press, Berkeley and Los Angeles.

Talling, J.F. (1961). Photosynthesis under natural conditions. *Annual Review of Plant Physiology.* 12, 133-154.

Trainor, F.R. (1978). *Introductory Phycology.* J. Wiley and Sons.

Vollenweider, R.A. (1969). *Primary Production in Aquatic Environments,* IBP Handbook no. 12. Blackwell Scientific Publications, Oxford.

Werner, D. (1977). *The biology of diatoms.* Bot. Monogr. 13. Blackwell Scientific Publications. Oxford.

Wetzel, R.G. (1975). *Limnology.* Saunders, Philadelphia.

INDEX

T

U

V

W

Z